THE BEGINNING

Our Universe and Others

Martin Rees

Astronomer Royal.

SIMON & SCHUSTER
A VIACOM COMPANY

First published in Great Britain by Simon & Schuster Ltd, 1997
A Viacom Company

Simon & Schuster Ltd
West Garden Place
Kendal Street
London W2 2AQ

Simon & Schuster of Australia Pty Ltd
Sydney

A CIP catalogue record for this book is available from the British Library

ISBN 0-684-81766-7

Typeset in Amasis 11/13pt by
Palimpsest Book Production Limited, Polmont, Stirlingshire
Printed and bound in Great Britain by
Butler & Tanner Ltd, Frome and London

ACKNOWLEDGEMENTS

This book presents an individual view on cosmology – how we perceive our universe, what the current debates are about, and the scope and limits of our future knowledge. It would not have been finished, certainly not in its present style, without the stimulus of Nick Webb at Simon & Schuster. His editorial advice was invaluable: he urged, in particular, that I should speculate a bit, and include controversial topics that I might otherwise have shied away from. The book is intended for general readers who share Nick Webb's infectious fascination with how science is done, and with the fundamental questions that cosmologists are trying to tackle. I am also grateful to Richard Sword for the frontispiece, and to Catherine Reed and Ingrid Connell for editorial help.

CONTENTS

INTRODUCTION

Philosophy begins in wonder. And, at the end, when philosophic thought has done its best, the wonder remains.

A.N. Whitehead

THE COSMIC PERSPECTIVE

Our universe sprouted from an initial event, the 'big bang' or 'fireball'. It expanded and cooled; the intricate pattern of stars and galaxies we see around us emerged thousands of millions of years later; on at least one planet around at least one star, atoms have assembled into creatures complex enough to ponder how they evolved.

Cynics used to say, not unfairly, that there were just two facts in cosmology: that our universe is expanding, and that the sky is dark at night. But no longer. Telescopes, on the ground and in space, have surveyed galaxies so far away that their light set out when our universe was a tenth its present age. Computers can simulate how galaxies emerged from amorphous beginnings. Other techniques have revealed 'relics' from still earlier eras of cosmic history.

We can trace the evolution of our universe back to when it was only a second old. This claim would have astonished earlier generations of cosmologists, who viewed their subject as a mathematical exercise remote from empirical checks. I would bet at least ten to one that there was indeed a big bang: that everything in our observable universe started as a compressed fireball, far hotter than the centre of the Sun. Most cosmologists would offer equally strong odds. (There would, though, still be vocal dissent from a minority.)

COSMOS AND MICROWORLD

Colossal though they may be, stars and galaxies rank low on the scale of complexity. That is why it isn't overpresumptious to aspire to understand them. A frog poses a more daunting scientific challenge than a star.

Planetary systems are common around other stars. But, given a propitious environment, what's the chance of life getting started, and of evolving to an 'interesting' stage? This biological question is still unsettled. The cosmos could be teeming with life. On the other hand, organic evolution could entail such a rare combination of 'accidents' that only our Earth harbours conscious intelligent beings.

Our 'cosmic cycle' may be finite, fated to end in a universal collapse into a 'big crunch'. But this won't happen before the stars have faded, and all atoms and even black holes have been recycled into radiation. Even if life and intelligence were now unique to Earth, they would have time to spread through our entire Galaxy and beyond. If life on Earth were snuffed out now, the potentialities of our entire universe would be diminished. Our biosphere's significance may be universal, not 'merely' terrestrial.

We can confidently trace cosmic history back to the first second. The ground gets shakier when we extrapolate still further back, into the first millisecond. But recent progress brings previously speculative questions within the scope of serious enquiry: Why is our universe so large? Why, indeed, did it expand at all?

The challenge facing the next Newton or Einstein is to 'unify' the forces of nature: to interpret electric, nuclear and gravitational forces as different manifestations of a single primeval force. This unification could manifest itself (and be tested) only when energies are enormously high – only, perhaps, in the initial instants of the big bang, when everything astronomers can see was squeezed smaller than a golfball, and quantum waves could shake the entire fabric of space. The 'seeds' for galaxies and other cosmic structures, and the ethereal 'dark matter' that pervades our universe, are residues of that era.

THE MULTIVERSE

As our universe cooled, its specific mix of energy and radiation, even perhaps the number of dimensions in its space, may have arisen as 'accidentally' as the patterns in the ice when a lake freezes. The physical laws were themselves 'laid down' in the big bang.

Our universe, and the laws governing it, had to be (in a well-defined sense) rather special to allow our emergence. Stars had to form; the nuclear furnaces that keep them shining had to transmute pristine hydrogen into carbon, oxygen and iron atoms; a stable environment, and vast spans of space and time, were prerequisites for the complexities of life on Earth.

The apparent 'fine tuning' on which our existence depends could be a coincidence: I once thought so. But that view now seems too narrow. What's conventionally called 'the universe' could be just one member of an ensemble. Countless other universes may exist in which the laws are different. The universe in which we've emerged belongs to the unusual subset that permit complexity and consciousness to develop. Once we accept this, various apparently 'special' features of our universe – those that some theologians once adduced as evidence for providence or design – occasion no surprise. This line of thought – the enlarged perspective of the 'multiverse' – supplies a motive for this book.

This new concept is, potentially, as drastic an enlargement of our cosmic perspective as the shift from pre-Copernican ideas to the realisation that the Earth is orbiting a typical star on the edge of the Milky Way, itself just one galaxy among countless others. Cosmologists can now tackle, in a genuinely scientific spirit, a new range of fundamental issues that they could previously just speculate about in non-professional moments.

Our entire universe may be just one element – one 'atom', as it were – in an infinite ensemble: a cosmic archipelago. Each universe starts with its own big bang, acquires a distinctive imprint (and its individual physical laws) as it cools, and traces out its own cosmic cycle. The big bang that triggered our entire universe is, in this grander perspective, an infinitesimal part of an elaborate structure that extends far beyond the range of any telescopes.

Some cosmologists speculate that new 'embryo' universes can form within existing ones. Implosion to a colossal density (around, for instance, a small black hole) could trigger the expansion of a new spatial domain inaccessible to us. Universes could even be 'manufactured' – the experimental challenge is far beyond present human resources, but may become feasible, especially if we recall that our universe has most of its course still to run. No information could be exchanged with a daughter universe, but it could bear the imprint of its parentage. Our own universe might be the (planned or unplanned) outcome of such an event in some preceding cosmos. The traditional theological 'argument from design' then reasserts itself in novel guise.

Most naturally created universes would be 'stillborn' in the sense that they could not offer an environment propitious for complex evolution: they would have too short a timespan, the wrong number of dimensions, allow no chemistry or be otherwise maladjusted. But our universe may not be the most complex: others in the ensemble may have richer structure, beyond anything we can imagine.

We can't grasp the nature of our cosmic environment by thought alone. It is modern telescopes and spacecraft – probing deeper into space, further back in time, and seeking bizarre objects such as black holes and cosmic strings – that have made cosmology a science. This book describes some high points of that quest, emphasising discoveries and ideas that are only now coming into focus. I have, however, tried to set the historical context, and to clarify some 'old' issues that perennially come up in discussions – the nature of redshifts, dark matter, gravity, and so forth.

I've interspersed the text with impressions or reminiscences of some outstanding figures I've encountered or worked with, and how their approach is moulded by their personal style, extra-scientific attitudes and, sometimes, their obsessions.

Here is the entry under 'feather' in Merriam-Webster's *Collegiate Dictionary*:

> Any of the light horny epidermal outgrowths that form the external covering of the body of birds and that consist of a shaft bearing on each side a series of barbs which bear barbules which in turn bear barbicels commonly ending in hooked hamuli and interlocking with the barbules of an adjacent barb to link the barbs into a continuous vane.

This definition is, I suppose, helpful to someone of immense learning who has never seen a bird. Some books about science seem aimed, likewise, at an (almost non-existent) class of readers: those with massive vocabularies who are nonetheless baffled by numbers or formulae. Jargon can be even less penetrable than simple equations.

I've tried to avoid jargon and formulae. But one can't avoid numbers. And, as is inevitable when describing the cosmos, some of these numbers are very big. What's important is just their order of magnitude, not the exact values, so they are presented in power-of-ten notation (10^x where x denotes the number of zeros in the number if we write it out in full).

Niels Bohr, a great pioneer of modern physics, advised his colleagues to 'speak as clearly as you think, but no more so'. He certainly heeded his own advice – indeed, he was famous for mumbling inaudibly and incomprehensibly. The mathematics used by theorists, and the instruments used by observers, may indeed seem abstruse. But these *technicalities* needn't concern anyone but specialists: they are just means towards tackling the great questions of cosmology: How did stars, planets and life emerge? Why is our universe the way it is? What imprinted the laws that govern it? Could other universes exist? Everyone can ponder these questions – indeed, when we are all groping, the specialist has less of a 'head start' over the general inquirer.

Some assertions about our universe are supported by firm evidence, and command wide assent among cosmologists; others are very tentative or speculative. I've tried not to conflate the two. This book describes some ideas that are part of the current consensus, as well as conjectures that my colleagues wouldn't share. I have tried, though, to maintain the distinction between what is well established and what isn't – or at least isn't *yet*.

My Cambridge colleague, Stephen Hawking, claimed in *A Brief History of Time* that each equation he included would have halved the book's sales. He followed that injunction, and so have I. But he (or maybe his editor) judged that each mention of God would double the sales. In flattering imitation, God has figured in the title of several subsequent books – *The God Particle*, *The Mind of God* and suchlike. In that latter respect I shall not follow Stephen's lead. Scientists' incursions into theology or philosophy can be embarrassingly naïve or dogmatic. The implications of cosmology for these realms of thought may be profound, but diffidence prevents me from venturing into them. I concur, rather, with another colleague, the cosmologist Joseph Silk: 'Humility in the face of the persistent great unknowns is the true philosophy that modern physics has to offer.'

Martin Rees
Cambridge, 1996

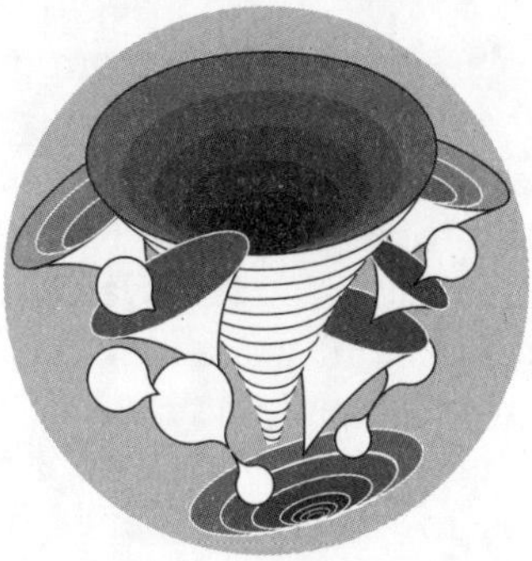

1

FROM ATOMS TO LIFE: GALACTIC ECOLOGY

> I am part of the sun as my eye is part of me. That I am part of the earth my feet know perfectly, and my blood is part of the sea.
>
> D.H. Lawrence

'WHILST THIS PLANET HAS BEEN cycling on according to the fixed law of gravity, from so simple a beginning forms most wonderful . . . have been and are being evolved'. These are the closing words of Charles Darwin's *Origin of Species.*

Cosmologists go back *before* Darwin's 'simple beginning' and aim to set our Solar System in a grand evolutionary scheme, stretching back to the emergence of the Milky Way Galaxy – right back, even, to the big bang that set our universe expanding. We are then emboldened to speculate: What are the potentialities of cosmic evolution in the far future? Could there be other universes, perhaps governed by different laws? If so, would they offer equally propitious environments for 'forms most wonderful' to evolve? These will be the theme of later chapters.

The Sun's light takes eight minutes travelling to Earth, and only a few hours to pass beyond Neptune and Pluto, the outermost planets of our Solar System. Light from the bright stars in the Milky Way – other suns like our own – has taken centuries to reach us. But even the entire Milky Way, the galaxy of stars that our Sun belongs to, is itself just a foreground feature in the cosmic vista. Our horizon now extends to objects so far away that their light set out several billion (thousand million) years ago.

Even if we knew nothing of the vast *spatial* scales revealed by modern telescopes, our Solar System stretches our conception of *time*scales to an extent that is hard to relate to human (or even historical) perspectives. Suppose America had existed for ever, and you were walking across it, starting on the East Coast when the Earth formed, and ending up in California when the Sun was about to die. To make this journey, you'd have to take *one step every two thousand years*. A mere three or four steps would represent all recorded history. Moreover, these few steps would be just before the halfway stage, somewhere in Kansas, perhaps – in no sense the journey's culmination. Our Sun is less than halfway through its life; we are still near the 'simple beginning' of the evolutionary story.

THE SUN AND ITS METABOLISM

The Sun and other stars are giant globes of incandescent gas. Inside them, two forces compete against each other: gravity and pressure. Gravity tries to pull everything towards the centre; but when gas is squeezed it heats up, and the resultant pressure balances gravity. The Sun's surface glows white-hot, at a temperature of 6000 degrees. But, to supply enough pressure, its centre must be far hotter still – about 15 million degrees, in fact.

What keeps the Sun shining? Without a source of fuel, gravity would gradually deflate the Sun as its heat leaked out. The great 19th-century Scottish physicist, Lord Kelvin, calculated that the

Sun would then shrink to half its present size in about 10 million years. Though a long time, this was not long enough: it fell far short of Darwin's favoured timespan for biological evolution, and of contemporary estimates of the Earth's age from geological strata and erosion. The Sun's life could be prolonged, Kelvin realised, only if there were 'some unknown source of energy laid down in the storehouse of creation'. That 'subatomic' energy must be implicated became clear in the 1920s: Einstein's famous equation $E = mc^2$ meant that energy was latent in all matter, and only a fraction of one per cent of the Sun's mass would suffice to keep the Sun shining. By the 1930s, enough had been learned about nuclear energy to resolve Kelvin's apparent paradox.

The Sun is fuelled by the same process that makes hydrogen bombs explode. Atoms of hydrogen are the simplest atoms of all: their nucleus consists of just one proton. The hotter a gas gets, the faster its constituent atoms move. In the Sun's core, protons crash together so hard that they stick. A series of reactions can fuse four nuclei of hydrogen (protons) into one helium nucleus. The helium nucleus, however, weighs 0.7 per cent less than the four hydrogens that made it. So fusing of hydrogen into helium releases $0.007mc^2$ – enough to keep the Sun shining for several billion years. The energy release in a star is steady and 'controlled', not explosive as in a bomb. This is because gravity pulls down the overlying layers firmly enough to 'hold the lid on', despite the huge pressure in the stellar core. The Sun has homeostatically adjusted so that fusion supplies power at just the rate needed to balance the heat shining from its surface, on which life on Earth depends.

The Sun is now understood, at least in broad outline. Current debate focuses on finer details: What causes the dark spots and turbulent flares on its surface? How can we learn about its deep interior? Does it shine steadily, or does its brightness fluctuate enough to affect the Earth's climate?

The Sun was born in an interstellar cloud. The cloud started with a barely perceptible rotation, but spun faster as it contracted (as does an ice skater who pulls in her arms); centrifugal force built up until it balanced gravity. A swirling disc then developed around a central proto-Sun which continued to deflate very gradually (more or less as Kelvin envisaged), but this slow contraction halted when the centre got hot enough to trigger hydrogen fusion. Meanwhile, the surrounding disc cooled; some of the gas condensed into dust and rocky fragments, which agglomerated into planets.

The Sun itself, by then orbited by a planetary system, settled into an equilibrium, slowly but steadily fusing hydrogen into helium. This can release so much heat that less than half the Sun's central hydrogen has so far been used up, even though it's already 4.5 billion years old. It will keep shining for a further 5 billion years. It will then swell up to become a red giant, large and bright enough to engulf the inner planets, and to vaporise all life on Earth. After this red-giant phase, some outer layers are blown off, leaving the core to contract into a white dwarf – a dense star no larger than the Earth, though several hundred thousand times heavier. This will shine with a blue glow, no brighter than the Moon today, on whatever then survives of our Solar System.

OTHER STARS

How does the brilliance and colour of a star depend on its mass, its age or on what it's made of? Astrophysicists can now answer such questions. They can compute, just as easily as the Sun's evolution, the life cycles of stars that start off lighter or heavier: half the Sun's mass, twice, four times and so on. Heavier stars are brighter, and trace out their life cycle more quickly. These computations use, as input, what physicists have learned about atoms and nuclei from laboratory experiments.

But how can such claims be tested? Stars live so long compared with astronomers that we're granted just a single 'snapshot' of each one's life. Nevertheless, we *can* test our theories, by surveying whole populations of stars. Trees can live for hundreds of years, but, even if you had never seen a tree before, it would take no more than an afternoon wandering around in a forest to deduce the life cycle of trees: you would see saplings, fully grown specimens and some that had died. William Herschel, the great 18th-century astronomer who discovered Uranus and surveyed the Milky Way, offered a picturesque metaphor:

> Is it not almost the same thing, whether we live successively to witness the germination, blooming ... and wither of a plant,

> or whether a vast number of specimens, selected from every stage through which the plant passes, be brought at once to our view?

Stars are easiest to observe when they are in the brightest phases of their evolution: well-known stars like Betelgeuse and Arcturus are in the 'giant' phase. The best 'test beds' for checking our theories of how stars evolve are the so-called globular clusters – swarms of up to a million stars, of different sizes, held together by their mutual gravity, which formed at the same time.

White dwarfs, the 'cinders' left behind when stars like the Sun complete their life cycles, are very common in our Galaxy, but because they are so faint it is harder to study them. Those that are freshly formed have very hot surfaces (and are actually blue rather than white); but they gradually cool, because they cannot draw on any further nuclear energy to compensate for what is radiated away. We can infer the temperature of white dwarfs from their colour (they redden as they cool), and theory then tells us their age (that is to say, the time that has elapsed since their parent stars exhausted their main nuclear fuel). The coolest are several billion years old, and this in itself tells us that some stars had completed their life cycles before our Solar System came into existence.

VIOLENT DEATHS

Not everything in the cosmos happens slowly. Sometimes stars explode catastrophically as supernovae. A nearby supernova flares up for a few weeks to be far brighter than anything else in the night sky. The most famous such event was observed in China. 'On a chi-chhou day in the fifth month of the first year of the Chih-Ho reign period' (July AD 1054), Yang Wei-Te, the Chief Computer of the Calendar – the ancient Chinese counterpart, perhaps, of the English Astronomer Royal – addressed his Emperor in these deferential words: 'Prostrating myself before your majesty, I have observed the appearance of a guest-star. On the star there was a slightly iridescent yellow colour.'

Now, nearly a thousand years later, we see the debris from the explosion witnessed by Yang Wei-Te – a bluish oval, containing filamentary gas expanding from a common centre. It's called the Crab Nebula. It will remain visible, gradually expanding and fading, for a few millennia; it will then become so diffuse that it merges with the very dilute gas and dust that pervades interstellar space.

The closest supernova of the 20th century – not as close as the Crab Nebula, but bright enough to be studied in detail – was seen in 1987. On the night of 23–24 February, Ian Shelton, a Canadian astronomer observing in the Chilean Andes, noticed a 'new' feature in the southern sky, in a concentration of stars known as the Large Magellanic Cloud. Its brightening and gradual fading were followed not only by optical astronomers but by those using the other modern techniques – radio, X-ray and gamma-ray telescopes – which have opened new 'windows' on the universe. Theorists had a rare and lucky chance to check their elaborate computations.

COSMIC ALCHEMY

These events fascinate astronomers. But why should anyone else care about exploding stars thousands of light-years away? Why should they interest the 99.9 per cent of people whose professional concerns are terrestrial rather than cosmic? Because, were it not for supernovae, the complexities of life on Earth could never have emerged – and we certainly wouldn't be here.

Ninety-two different kinds of atom occur naturally on Earth, but some are vastly more common than others. For every 10 atoms of carbon, you'd find, on average, 20 of oxygen, and about 5 each of nitrogen and iron. But gold is a hundred million times rarer than oxygen, and others – uranium, for instance – are rarer still.

Everything that's ever been written in our language is made from an alphabet of just 26 letters. Likewise, atoms can be combined in a huge number of different ways into molecules: some as simple as water (H_2O) or carbon dioxide (CO_2), others containing thousands of atoms. The most important ingredients of living things (ourselves included) are carbon and oxygen atoms, linked (along with others) into chainlike molecules of huge complexity. We couldn't exist if these particular atoms weren't common on Earth.

Atoms are themselves made up of simpler particles. Each kind has a specific number of protons (with positive electric charge) in its nucleus, and an equal number of electrons (with negative electric charge) orbiting around it: this is called the atomic number. Hydrogen is number 1; uranium is number 92.

The nuclei of all atoms are made up of the same elementary particles – protons, together with neutrons – so it's not surprising that they can be changed into one another. This happens, for instance, in a nuclear explosion, but nuclei are so 'robust' that they aren't themselves destroyed by the chemical changes that occur in living things, or in laboratories.

The different kinds of atom on Earth exist in the same proportions as when the Solar System formed about 4.5 billion years ago: no natural terrestrial process can create or destroy the atoms themselves.[1] We'd like to understand why they were 'dealt out' in these particular proportions. We could leave it at that – perhaps a creator turned 92 different knobs. But it's natural to seek a less 'ad hoc' explanation, and try to trace complex structures back to simple beginnings. In this instance astronomers have supplied the key insights: it seems that the universe indeed started out with simple atoms, which were fused and transmuted into the heavier ones inside stars.

Not even the centre of the Sun is hot enough to perform these transmutations. But the cores of bright blue stars like those in the Orion nebula, and the intense shocks when they finally explode, can transmute base metals into gold.

Stars more than ten times heavier than the Sun shine more brightly; they evolve in a more complicated and dramatic way. Their central hydrogen gets consumed (and turned into helium) within a hundred million years – less than one per cent of the Sun's lifetime. Gravity then squeezes these heavy stars further, and their central temperature rises still higher, until helium atoms can themselves stick together to make the nuclei of heavier atoms – carbon (six protons), oxygen (eight protons), and iron (26 protons). A kind of 'onion skin' structure develops: a layer of carbon surrounds one of oxygen, which in turn surrounds a layer of silicon. The hotter inner layers have been transmuted further up the periodic table and surround a core that is mainly iron.

When its fuel has all been consumed (in other words, when its hot centre is transmuted into iron), a big star faces a crisis. Catastrophic infall compresses its core to the density of an atomic nucleus, triggering a colossal explosion that blows off the outer layers at

10 000 kilometres per second. This explosion manifests itself as a supernova of the kind that created the Crab Nebula. The debris contains the outcome of all the nuclear alchemy that kept the star shining over its entire lifetime. There is a lot of oxygen and carbon in this mixture, plus traces of many other elements. The calculated 'mix' is gratifyingly close to the proportions now observed in our Solar System.

SOME HISTORY

It was Hans Bethe, one of the leading pioneers of nuclear physics in the 1930s, who first elucidated how energy is released inside the Sun.[2] Accurate calculations of what happens in stellar cores, especially during the later hotter phases that precede the supernova outburst, involve very complex reaction networks, and have had to await powerful computers. The outcome depends rather delicately on the detailed nuclear physics. (Some of the techniques needed for these intricate computations were developed by specialists in nuclear weapon design. It is therefore not surprising that the first detailed supernova calculations emerged from the US Livermore Laboratory and similar establishments, and from their Soviet counterparts.) Hans Bethe's career now spans nearly 70 years. He was head of the theoretical division at Los Alamos, during the development of the first atomic bomb; throughout the last half-century he has been an untiring advocate of arms control. But he remains, remarkably, at the forefront of supernova theory.

How the stars relate to the elements around us – why carbon and iron atoms are common, but gold atoms are rare – was first clearly conceived by Fred Hoyle, in spare moments when he was involved in developing radar during the Second World War.

Had Fred Hoyle been born ten years earlier, he might have shared in the triumphant achievements of the 'heroic age' of theoretical physics between 1925 and 1930. During those few years, the quantum theory was formulated, bringing order into the seemingly paradoxical properties of atoms, electrons and radiation. All scientists, whether they are cosmologists or biologists, must surely concede that quantum mechanics surpasses any other conceptual breakthrough in the breadth of its scientific ramifications, and in the jolt its 'counterintuitive' consequences gave to our view of nature; the microworld is fully as strange as the cosmos. But no single 'Einstein figure' was responsible; instead, a brilliant cohort – Erwin Schrödinger, Werner Heisenberg and Paul Dirac – laid the foundations of this new world-view.

The late 1930s were a period of consolidation after the revolutionary new ideas of the 1920s. Dirac, by then a professor at Cambridge, advised Hoyle, then recently graduated, that 'In 1926, people who were not very good could do important work in basic physics. Today [1938], even people who *are* very good can't find important problems to solve.' So Hoyle shifted his attention to the stars. He applied what he knew about atomic nuclei, to conjecture how they might behave at ultra-high temperatures.

Hoyle knew that the heavier atoms, higher up the periodic table, tend to be sparser on Earth. Magnesium and silicon are less abundant than oxygen; and the precious metals are a million times rarer. But there is an exception to this general trend: the 26th element, iron, is relatively common, and so are its neighbours in the periodic table. Hoyle also knew, from his basic physics, that iron and its neighbours are the nuclei with the highest 'binding energy': they are the most stable, and an input of energy is needed either to break them up or to transmute them into still heavier nuclei. So could the different chemical elements be the outcome of nuclear transmutation? Even without knowing the detailed reaction network, one can confidently infer that iron peak nuclei, once formed, would be harder to destroy. So the relative abundances would end up with an 'iron peak'.

But such 'processing' requires an environment where the nuclei can actually be transmuted into one another. This is a more demanding requirement for the heavier nuclei: they repel each other more strongly because of their bigger electric charges (iron carries the charge of 26 protons); to merge or disrupt them therefore requires much more energetic impacts than are needed, for instance, to convert hydrogen (one charge) into helium (two charges). The random speed of atoms in a gas depends on its temperature, so this transmutation implies extreme temperatures. Hydrogen fusion in the Sun requires 15 million degrees. But Hoyle estimated that the environment where the 'iron peak' elements were forged must have been a hundred times hotter still – more than a *billion* degrees.

Hoyle published his prescient conjectures in 1946. All the 'heavy' elements on Earth were, he claimed, built up from simpler nuclei inside stars that, during their later evolution, reached billion-degree temperatures. Stars like the Sun never get as hot as Hoyle envisaged. But perhaps more massive stars do?

Hoyle was encouraged in his original ideas by the evidence of abundances on Earth. But are these in any sense 'typical' of the cosmos? In one respect they are not. Hydrogen and helium would have been too volatile to have been effectively retained in the proto-Earth; so these two lightest elements (which we now know to be far the most abundant in the Sun), are underrepresented on Earth. However, the proportions of the others may be more typical of the Solar System.

But what about the rest of our universe? What are the stars made of? The French philosopher Auguste Comte, 150 years ago, averred that this would always be unanswerable. In his *Cours de Philosophie Positive* he wrote, 'We shall never be able to study, by any method, their chemical composition or their mineralogical structure . . . Our positive knowledge of stars is necessarily limited to their geometric and mechanical phenomena.' But, even before the 19th century was over, astronomers had appreciated the rich information carried by starlight. When the light passes through a prism, and disperses into a spectrum, we see the telltale colours of different substances – oxygen, sodium, carbon and the rest. The element helium, number two in the periodic table, was not recognised on Earth until its distinctive spectral features had been noted in the Sun's spectrum. Stars are made of the same kinds of atoms as we find on Earth. But it has proved a more subtle task – and one that still engages many astrophysicists – to use these spectra to infer how abundant the various atoms are in different stars and nebulae.

Fleshing out Hoyle's ideas required better data on the nuclear reactions that played an especially critical role; Hoyle managed to inspire the enthusiasm of W.A. (Willy) Fowler, a physicist at Caltech, who focused his laboratory's efforts on astronomically interesting measurements. The whole scheme of cosmic nucleogenesis, as it stood in 1957, was codified by Hoyle and Fowler in a book-length article co-authored with their colleagues, Geoffrey and Margaret Burbidge. This classic article – known to all astronomers as 'B^2FH', the initials of its four authors – has stood the test of time.

The most important advance since 'B^2FH' concerns the elements further up the periodic table than iron – because iron is the most tightly bound nucleus, energy has to be *supplied*, rather than being released, in creating still heavier nuclei like lead and uranium. The key ideas here came from Hoyle's younger American collaborators, who worked together with him during regular summer visits to his Institute in Cambridge: a process called 'explosive nucleosynthesis' occurs during the supernova explosion itself, when material gets suddenly heated by a shock that blasts its way outward through the star.

THE ECOLOGY OF THE MILKY WAY

Primo Levi, in his book *The Periodic Table*, depicts the eventful history of a typical carbon atom on the Earth:

> It lies bound to three atoms of oxygen and one of calcium, in the form of limestone. Then it was cut out, went through a lime furnace, and took wing. The atom was caught by the wind, flung down onto the earth, lifted ten kilometres high. It was breathed in by a falcon . . . dissolved three times in the sea, and again was expelled. It then stumbled into organic adventures. It had the good fortune to brush against a leaf, penetrate it, and be nailed there by a ray of the Sun.
>
> In an instant, like an insect caught by a spider, the carbon atom was separated from its oxygen, combined with hydrogen, and finally inserted in a chain of life.
>
> . . . It enters the bloodstream, knocks at the door of a nerve cell, enters, and supplants the carbon that was part of it. The cell belongs to a brain and it is my brain; the cell in question, and within it the atom in question, are in charge of my writing, in a mysterious game that nobody has yet described. It . . . guides this hand of mine to impress on this paper this dot here, *this* one.

The theory of stellar evolution and nucleogenesis, an undoubted triumph of 20th-century astrophysics, extends each atom's history back before the Earth formed. A galaxy resembles a vast ecological system. Pristine hydrogen is transmuted, inside stars, into the basic building blocks of life – carbon, oxygen, iron and the rest. Some of this material returns to interstellar space, thereafter to be recycled into new generations of stars.

Our Galaxy, the Milky Way, is a huge disc 100 000 light-years across and containing a hundred billion stars. Its oldest stars formed more than ten billion years ago. The primordial material contained only the simplest atoms – no carbon, no oxygen and no iron. Our Sun is a middle-aged star: some others are more than twice as old. Before it even formed, 4.5 billion years ago, several generations of heavy stars could have been through their entire life cycles. The chemically interesting atoms – those essential for complexity and life – were forged inside these stars. Their death-throes, supernova explosions, flung these atoms back into interstellar space.

A carbon atom, forged from helium in an early supernova, might have wandered for hundreds of millions of years between the stars. It might then have found itself in an interstellar cloud, which collapsed under its own gravity to form stars. The atom might have entered the core of some new bright star, and been processed further up the periodic table (into silicon or iron), and then been flung back in another supernova. Or it might have joined a less massive star, surrounded by a spinning gaseous disc that condensed into a retinue of planets. One such star could have been our Sun. This carbon atom might have found itself in the newly forming Earth, to play its part in the geological processes that moulded and weathered the Earth's surface; and in the chemistry whereby species emerged and evolved, ending, perhaps, in Primo Levi's brain-cell.

Despite the regularities in their proportions relative to *each other*, the amounts of carbon, oxygen, sodium and the other 'heavy elements' *relative to hydrogen* are not the same everywhere. Their abundances are *lower* in the *oldest* stars. This is of course just what would be expected if they had indeed been synthesised gradually, by successive generations of stars. The oldest stars, forming early in Galactic history, would have condensed from material that had not yet become as heavily polluted as it is today (and as it was when younger stars formed). A second trend is that the abundances are higher in locations where star formation is fastest and the recycling is more efficient.

In our Galaxy's youth, there was no carbon, oxygen or iron, and chemistry would have been a dull subject. Before complex chemical compounds could form, and before a Solar System could emerge, ancient stars had to do the basic work of synthesising, transmuting and recycling the chemical elements.

Carbon atoms – those in every cell of your blood or your brain, or in the ink on this page – have a pedigree extending back far earlier than our Solar System's birth 4.5 billion years ago. The Solar System itself condensed from the intermingled debris of many earlier stars. If we trace the histories of the atoms further back – back to (say) seven billion years ago – we'd find them spread through the entire Galaxy. Atoms now linked together in a single strand of DNA were then inside different stars spread around the Galaxy, or dispersed in the medium between those stars.

In the distant future, after the death of our Solar System, these atoms may disperse once more through the Galaxy, and be incorporated into new stars. Their entrapment in the same DNA molecule is a transient biochemical event; likewise, in astrophysical perspective, their entrapment in the same Solar System is just a transient phase in a history that stretches back before the Galaxy formed, and may extend into an infinite future.

Why are carbon and oxygen atoms so common here on Earth, but gold and uranium so rare? This everyday question isn't unanswerable – but the answer involves ancient stars that exploded in our Milky Way more than five billion years ago, before our Solar System formed. The cosmos is a unity. To understand ourselves we must understand the stars. We are stardust – the ashes from long-dead stars.

OTHER PLANETS?

Stars are still forming today. About 1500 light-years away lies the Orion Nebula: enough gas and dust to make millions of stars. It contains bright young stars; it even contains protostars that are still condensing, and haven't yet become hot enough to ignite their nuclear fuel. Spinning discs of dusty gas encircle some protostars. These are proto-Solar Systems: dust particles stick together to make rocky 'planetesimals', which merge to make planets.

Planetary systems used to be attributed to improbable and unusual events – one idea, for instance, was that another star passed so close to the Sun that its gravity tore out a stream of gas, which cooled to form the planets. But it is now clear that planet formation does not require any rare 'accident'. Planets are a natural concomitant of star formation. Indeed, they are inevitable unless the material that formed a star happened to have essentially zero spin, and it is *that* which would involve a rare coincidence.

Planetary systems should therefore be widespread. Fully formed planets orbiting other stars would look very faint. However, their effect could be discerned indirectly. A star and its attendant planets trace out orbits which pivot around their fixed common centre of mass, called the 'barycentre'. The barycentre is, of course, much closer to the star than to the planets, because the star is much heavier. The star is therefore displaced only slightly; but sufficiently precise measurements might detect the small motions induced by orbiting planets.

The first really convincing evidence for a planet around an *ordinary* star[3] didn't come until 1995, when Michel Mayor and David Queloz of the Geneva Observatory found that the Doppler shift of 51 Pegasus, a star resembling our Sun and lying 40 light-years away, was going up and down very slightly every four days; a planet almost as heavy as Jupiter seems to be moving in a very close orbit around it. This planet would be ten times closer to its parent star than Mercury is to our Sun, and its surface would be hotter than 1200 degrees Celsius; but it may be just the largest member of another entire solar system. Within a few months Geoffrey Marcy and Paul Butler, in California, had discovered planets around other stars: these have slower orbits, and would be at the right temperature for water to exist. But these planets are all very large – heavier than Jupiter. Planets weighing no more than the Earth would be a hundred times harder to detect.

Planets on which life could evolve, as it did here on Earth, must be rather special. Their gravity must pull strongly enough to prevent the atmosphere from evaporating into space; they must be neither too hot nor too cold, and therefore the right distance from a long-lived and stable star. Only a small proportion of planets meet these conditions, but planetary systems are (we believe) so common in our Galaxy that Earthlike planets would be numbered in millions. The policy-makers of NASA have urged that a search for planets should become a main thrust of the US space programme: the goal is of immense scientific interest, and would be more likely than most other scientific programmes to fire public enthusiasm.

The task is difficult not only because an Earthlike planet would be so faint, but because the planet and star would lie so close together on the sky that no existing optical telescope, even the Hubble Space Telescope, yields sharp enough images to separate the two. The technique of 'interferometry' must be employed, whereby two separate telescopes are linked. In contrast, the Jupiter-like planets were discovered rather cheaply, using modest-sized telescopes on the ground equipped with ingenious equipment for analysing the starlight so as to reveal the very slight motions of the star induced by an orbiting planet.

The technical challenge of actually detecting Earthlike planets is nevertheless *not* insurmountable, and, once a candidate had been seen, several things could be learned about it. Suppose an astronomer forty light-years away had detected our Earth – it would be, in Carl Sagan's phrase, a 'pale-blue dot', very close to a star (our Sun) that outshines it by many millions. If Earth could be seen at all, its light could be analysed, and would reveal that it had been transformed (and oxygenated) by a biosphere. The shade of blue would be slightly different, depending on whether the Pacific Ocean or the Eurasian land mass was in view. Distant astronomers could therefore, by repeated observation, conclude that the Earth was spinning, and learn the length of its day, and even infer something of its topography and climate.

LIFE?

How life might emerge on a planet, even given the right physical environment, raises more subtle issues. 'Life' may take forms that we wouldn't recognise and can't conceive of. However, the probability that life of the terrestrial variety emerges elsewhere depends on the answers to two questions. First, how frequently do Earthlike environments occur? Second, what is the chance that life develops *even when* physical conditions are optimal?

The first of these questions has already been addressed: 'suitable' planets should be common. But even when a planet's chemistry, temperature and gravity are propitious, how likely is life to begin? There is still no consensus among biologists, but a tantalising clue emerged in 1996, when tiny traces of seemingly organic 'fossil' material were found in a meteorite believed to have been knocked off Mars. If life really emerged independently on our neighbouring planet, then the odds against igniting the 'green fuse' in any similar environment would shorten dramatically. But Martian life, even if it once existed, plainly stagnated at a primitive level. Our Earth's complex biosphere could have emerged very much against the odds; and the evolution of intelligence could have been even more of a fluke. The whole course of evolution has been channelled by 'random' events – cometary impacts, extinctions and the like.

If evolution on Earth could be rerun, the outcome might be quite different; biologists argue that there would still be animals with eyes, because the evolutionary record reveals that *eyes*, in some form or other, evolved independently many times. But would *intelligence* necessarily emerge? The evolutionist Ernst Mayr argues that intelligence is not (like eyes) a generic outcome of evolution because it seems to have emerged only once. But there may be another reason why intelligence could not evolve by several independent routes. Once intelligence has emerged, and developed beyond some threshold, it controls the biosphere, and natural selection no longer has free rein. Unless the dominant species wiped itself out, intelligence would never have a second independent chance to evolve.

Even when simple life exists, we don't know the chances that it evolves towards intelligence, nor how long it persists even when it does emerge. Intelligent life could be 'natural'; or it could have involved a chain of accidents so surpassingly rare that nothing remotely like it has happened anywhere else in our Galaxy.

WHY INTELLIGENT LIFE COULD BE RARE

There are actually quite cogent reasons to suspect that intelligent life is rare. One old argument is prompted by the great physicist Enrico Fermi's question, 'Where are they?' Many stars are several billion years older than our Sun, so evolution on other sites should have had a head start over us. Why, then, haven't aliens visited us, or at least created signals or artefacts that betray their presence clearly?

A quite different argument that advanced life is rare comes from Brandon Carter, an expert on black holes who will figure in later chapters. His starting-point is the well-known fact, already mentioned, that our Sun is about halfway through its life. In other words, the time we have taken to evolve is (within a factor of 2) the same as the Sun's age. Carter thought it strange that these two times came out roughly equal. Humans are the evolutionary outcome of an immense number of generations of the successive species that were our evolutionary precursors. The Sun's age is fixed by quite different (and much better-understood) physical constraints. These two timescales could, a priori, have differed by many powers of ten.

Carter offered a new perspective on these timescales. His argument goes like this. It would seem an unlikely coincidence that the typical time for emergence of intelligence should be the same as the lifetime of a star: the processes determining these two timescales are quite unrelated. The typical evolutionary timescale would (one might have guessed) have been either much shorter or much longer than the Sun's age. If it were much shorter, then we would be laggards – Fermi's question would have to be confronted squarely. On the other hand, there is no problem with the hypothesis that the biological timescale is *typically much longer* than stellar ages. In typical cases, evolution on most planets would not get very far before their parent star died. We wouldn't be here at all unless, here on Earth, the key steps in evolution had all happened especially quickly. Intelligent life, of the kind that evolves on planets around stars, should therefore be rare.[4]

BRIDGING COSMIC CULTURAL GAPS

All we can safely conclude is that intelligent life has evolved at least once. Even if it existed elsewhere, we might not recognise it. Intelligent extraterrestrials may lead purely contemplative lives, and have no motive for signalling their presence to us: absence of evidence cannot be evidence of absence.

Systematic scans for artificial signals are a worthwhile gamble – despite the heavy odds against success – because of the philosophical import of *any* detection. External intelligences could be 'organic' life; they could equally well be machines constructed by (or evolved from) such life. Either way, the 'conventional wisdom' is that they would most likely reveal themselves via *radio*-frequency signals – radio telescopes are exceedingly sensitive, and less power is needed to transmit detectable signals in the radio band than in (for instance) optical or X-ray pulses. Searches have concentrated on the so-called 'waterhole' waveband, between the 21cm wavelength emitted by atomic hydrogen (H) and the 18cm wavelength of OH (which, of course, together with another H atom, makes a water molecule).

It would be easy to devise signals that would be incontrovertably artificial: for instance, attention could be attracted by a series 1, 3, 5, 7, 11, 13, 17, 19, 23, 29 . . . These are the prime numbers: no natural process could generate them, but they would be recognised by any culture that was interested in (and capable of) picking up cosmic radio waves.

A series of carefully contrived messages could then build up a vocabulary for describing matters that might be common to sender and any possible receivers: basic mathematics, physics and astronomy. The logician Hans Freudenthal wrote an entire book entitled *Lincos: design of a language for cosmic discourse*, spelling out in detail how this might be done. Signals could also transmit images, and (a common science-fiction theme) instructions for replicating three-dimensional structures could be 'beamed down' to us. The nearest potential sites are so far away that signals would take many years in transit. For this reason alone, quite apart from the 'culture gap', transmission would be primarily one-way – there would be time to send a measured response, but no scope for quick repartee.

Optimists claim that such signals could convey enlightening messages of such import that they would enable us to bypass centuries of scientific endeavour and discovery, or be forewarned of (and steered away from) potential catastrophe. But such a gap would be hard to bridge, even within human culture. Could, for instance, a short 'message from the future' have guided a leading intellect from an earlier era towards some aspect of modern scientific knowledge? Could Newton have been steered from alchemy towards chemistry by, for instance, being urged to look through his prism at the spectra of flames made by burning different substances? Could Aristotle have been led towards more modern ideas in astronomy or anatomy? Many similar questions could easily be posed. It would be a daunting challenge to bridge even a few centuries of human cultural change, essentially because scientific advance depends on *gradual* advances of interconnected techniques and technologies.

The culture gap with extraterrestrials may be unbridgeable. But the *mere receipt* of an artificial signal would in itself have greater import than any hints or warnings it could convey about our shared environment. We would know that our Earth wasn't the only place where something 'interesting' had evolved, and that concepts of logic and physics weren't peculiar to the 'hardware' in human skulls.

Radio telescopes in several countries have been used to scan the sky for artificial transmissions. But even these small-scale efforts have had a hard time getting public funding because the topic is encumbered by manifestly 'flakey' connotations – UFOs and so forth. But there is surely more widely diffused public interest in this quest than in any traditional branch of physics or astronomy. If I were an American scientist testifying to a congressional committee, I would feel more confident and comfortable requesting 10 million dollars (less than some science-fiction movies gross in a single week) to search for extraterrestrial intelligence (SETI) than seeking funds for a 10-*billion*-dollar accelerator to probe subatomic physics.

The odds may be stacked so heavily against intelligent life that none has developed at any other site within our Galaxy – conceivably there is none anywhere in the part of the universe that we can observe. Some may find it depressing to feel alone in a vast mindless cosmos. But I would personally react in quite the opposite way. It would in some ways be disappointing if SETI searches were doomed to fail, but we could then envisage our Earth in a less humble cosmic perspective than it would merit if our universe already teemed with advanced life-forms.

A COSMIC COINCIDENCE

For life to emerge, the local conditions must be 'right'; but the entire universe must be propitious as well. The physical laws must allow atoms to combine into complex molecules and stable Solar Systems to form, and remain stable. There must be sufficient expanses of space and time for stars to evolve, and for their nuclear waste to be recycled into a new generation of stars, some with attendant planets. These are stringent demands: they would not be fulfilled in a 'typical' universe; as discussed later in this book, they hold clues to the origin of our universe, and perhaps of others.

Fred Hoyle was the first to discover one specific requirement of this type. He realised that the atomic 'building blocks' for life could be made in stars, but that the requisite transmutations happened only because atomic nuclei have rather special properties that would seem to a nuclear physicist to be no more than a surprising 'accident'.

In the early 1950s Hoyle was pondering how stars could create atoms of carbon and oxygen – elements that are abundant in the cosmos, and of course crucial for life. A carbon nucleus, with six protons and six neutrons, is made from three helium nuclei (each with two protons and two neutrons). Three helium nuclei would be unlikely to crash together simultaneously, even in the dense core of a star. Far more likely would be two successive 'two-body' collisions: the first would form beryllium, whose nucleus contains four protons and four neutrons; another helium could then be captured later, in a separate collision, to create carbon.

But there seemed a fatal problem with this simple picture. Beryllium is unstable. Its lifetime, known from laboratory measurements, was so short that there seemed little chance of a beryllium nucleus capturing a third helium nucleus in the brief interval before it decayed. There was no way through this bottleneck unless beryllium and helium could stick together especially easily and quickly. Hoyle realised that this could happen only if there happened to be a so-called 'resonance' in the carbon nucleus, whose energy matched that of the incoming beryllium and helium nuclei: the requirement, to state it more precisely, was that the total energy (mc^2) of a carbon nucleus in this resonant state should equal the energies of the two incoming nuclei added together, plus the kinetic energy of their impact.

Carbon nuclei had not yet, in the early 1950s, been fully investigated. Hoyle had recently begun his fruitful collaboration with the Californian nuclear physicist W.A. Fowler. He cajoled Fowler and his colleagues into checking whether carbon nuclei could behave as he predicted. Experiments indeed revealed a 'resonance', consistent with Hoyle's expectations. The energies of particles are measured in units called electronvolts – in chemical reactions, a few electronvolts (at most) are released by each atom. Nuclear reactions are millions of times more energetic than chemical reactions, so the usual unit is an MeV – a million electonvolts. Hoyle had predicted a 'resonance' in the carbon nucleus with an energy of 7.7 MeV; the experimenters found 7.65 MeV. We need not bother about how these energies are measured – the important point is that the two numbers are remarkably close to each other.

Without this particular resonance in its nucleus, carbon could not have been made in stars. For carbon to survive, there is another requirement: it must not too quickly capture a fourth helium nucleus, which would turn it into oxygen. This latter reaction, however, *isn't* especially efficient. But if the oxygen nucleus were only one per cent different, it too would have a resonance, and the carbon would be processed into oxygen, or even further up the periodic table, as quickly as it was made.

These features of carbon and oxygen are unremarkable in themselves. All nuclei have resonances, and these particular energies are no more or less likely than other values in the same general range. But, if you tried to guess a number on a lottery ticket that had an equal chance of being anything between (say) 1 and 1000, you would generally be wrong by at least 100; there is only about a two-per-cent chance that you'd guess a number closer than 10 to the the right answer. Hoyle's guess was so close that a nuclear physicist might have offered odds of 100 to one against him. These features of carbon and oxygen, seeming accidents of nuclear physics, turn out crucial for the pervasiveness of carbon in stars and planets, and therefore for the course of cosmic evolution.

The properties of atoms and nuclei depend on some numbers that are still more basic: the masses of the 'elementary' particles they are made of, and the strengths of the forces that bind them together. Hoyle speculated that these numbers were not truly universal, but might take different values in different regions. Complexity (and perhaps life) would then be confined to 'cosmic oases' where conditions were propitious – where, for instance, carbon had the unusual property that he had identified. But we now know enough about remote parts of our universe to be confident that the basic laws are the same everywhere our telescopes can probe.

Is it then simply a coincidence – a brute fact – that carbon nuclei have this special feature, even though the odds against it would seem like 100:1? Most cosmologists, for the last 30 years, would have answered 'yes'. But Hoyle could, in essence, have been right: his only error was not to think on a grand enough scale. Our *entire observable universe* could be an 'oasis' in a grand ensemble of other universes. Although we cannot observe them (and they may be for ever inaccessible) other universes are a natural expectation from current cosmology. Moreover, many features of our universe that otherwise seem baffling fall into place once we recognise this. A main theme of this book will be to elaborate the concept of a 'multiverse'. But, before speculating further, we must contemplate the scale and fabric of the universe that is our home.

NOTES

1 An exception should be made for the rare radioactive elements, which transmute spontaneously: for example, uranium decays slowly into lead. The best estimates of the Earth's age come from measuring the fraction of uranium that survives from the time the Earth first solidified

2 Hydrogen nuclei (protons) can combine into helium nuclei in two different ways. One involves direct 'proton–proton' reactions. The other (which can operate when some heavier nuclei are already present) is the so-called 'CNO cycle', which uses carbon, nitrogen and oxygen as catalysts, but has no net effect on the abundance of these elements.

3 As mentioned later in Chapter 4, a planetary system of an unusual kind had been discovered in 1992. Its central star is very different from the Sun: a spinning neutron star. This system would be an unpropitious abode for life.

4 In the same Royal Society lecture in which he presented this view, Carter adumbrated his even more controversial 'doomsday' argument, which purports to show that our species is unlikely to last more than a few more centuries. The argument is based on a (debatable) analogy with the following simple line of reasoning. Suppose a box contains *N* tickets, numbered from one to *N*, but you know nothing about how big *N* is: there could be 10 tickets in the box, there could be billions, or any number in between. You draw out a ticket at random, and the number on it is 2452. You might then guess that *N* was most likely around 5000, so that your choice was roughly halfway between the beginning and end of the sequence. You would, indeed, be slightly surprised if your number was in the first 5 per cent or the last 5 per cent of those represented in the box: there is a 90-per-cent chance that it is in neither of these categories. Therefore, having chosen 2452, you might have 90-per-cent confidence that *N* is neither less than about 2600 (otherwise you'd be in the last 5 per cent) nor more than 50 000 (which would mean you had drawn a ticket in the first 5 per cent). The 'doomsday' argument applies similar reasoning to the 'roll call' of all humans who have ever lived or will ever live in future. (Even though there have been tens of thousands of past generations of humans, the recent population rises have been so drastic that more than 10 per cent of those who have ever lived are alive today.) Carter concludes that the human population must terminate, or at least decline drastically, within a few centuries: we would otherwise come improbably early to the 'roll call.' This argument has been taken up, and

enthusiastically elaborated, by the Canadian philosopher John Leslie, and by Richard Gott (whose ingenious ideas on 'time machines' are mentioned in Chapter 13). I personally attach some weight to Carter's argument on the rareness of life in our universe, though I hope biologists will soon give us a firmer scientific assessment. But I can't take the 'doomsday' argument seriously, even though its depressing conclusion is not in itself implausible. This is partly because the predetermined value of N in the box of tickets doesn't seem a strict analogy for a number that depends on the indeterminate and 'open-ended' future, and could be infinite.

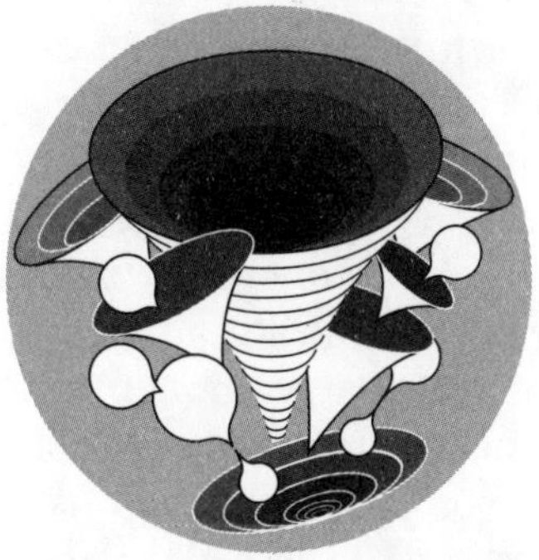

2

THE COSMIC SCENE: EXPANDING HORIZONS

> Yet it is possible that some bodies, of a nature altogether new, and whose discovery may tend in future to disclose the most important secrets in the system of the Universe, may be concealed under the appearance of very minute single stars no way distinguishable from others of a less interesting character, but by the test of careful and often repeated observations.
>
> John Herschel (1820)

No biologist would observe just one rat, and then grandly generalise about animal behaviour – individual rats might have 'hangups' peculiar to themselves. Nor would physicists happily base a theory on a single unrepeatable experiment. But we cannot directly check our ideas about our universe by looking beyond it. Despite these limitations, scientific cosmology *has* progressed, but only because our universe, in its large-scale structure, is simpler than we had any right to expect.

'Cosmology' originally meant a 'world-view' in the broadest sense. The word now connotes, more specifically, the study of the entire observable universe, treated as a single entity.

There turns out to be a clear demarcation between 'cosmology' and the rest of astronomy, which deals with the ingredients and fine-scale texture of our universe. A terrestrial analogy may clarify this. From a vantage point far out of sight of land, an ocean view reveals complex structures: waves (sometimes small riding on large), foam and so on. But, once your gaze extends beyond the longest ocean swells, you see an overall uniformity, stretching to the horizon many miles away. A piece of ocean large enough to be 'typical' must obviously extend several times further than the scale of the longest waves. But this is still small compared with the expanse of ocean we can see – our horizon extends far enough to encompass many regions statistically similar one to another, each large enough to constitute a 'fair sample'. This broad-brush uniformity of *seascapes* is not, however, a general feature of *landscapes*: on land, progressively larger mountain peaks may stretch to the horizon, and a single topographical feature may dominate the entire view.

Our observable universe – the volume out to the 'horizon' that powerful telescopes can reach – resembles a seascape rather than a mountain landscape. Even the most conspicuous features are small compared with the range of our telescopes. So it makes sense to talk about the average 'smoothed-out' properties of our observable universe. This characteristic, without which cosmologists would have made no progress, seemed until recently a gratifying contingency – only now are we coming to realise *why* our universe has this simplifying feature.

This metaphor can be pushed further. The ocean may seem uniform within our horizon, only a few miles distant. But that doesn't mean it extends featureless to infinity. A few hundred miles away, the weather may be much stormier or much calmer: the waves may look quite different. A thousand miles away, the ocean may be bounded by a shoreline. Likewise, our whole observable universe may be merely a patch of space and time in a richly structured multiverse. We observe symmetry and simplicity in our 'patch' because these structures are on scales vastly larger than we can directly probe.

Our observational horizon extends out to 10 billion light years, but encompasses only a fragment of physical reality – moreover, it may be a far from typical fragment. Beyond our horizon may lie new levels of complexity on far grander scales. I will speculate later that some enigmatic features of our universe – the 'coincidences' that make it an abode for life – can be explained only by broadening our conceptual horizon. Let us start this outward progression by considering the universe of galaxies beyond our own.

THE HIERARCHY OF COSMIC STRUCTURE: FROM STARS TO SUPERCLUSTERS

Our own Galaxy is typical of countless others scattered through our universe. Its constituent stars and gas lie mainly in a disc, spinning around a central 'bulge' where the stars are closer together and where (as we shall see in Chapter 5) there may lurk a huge black hole. A light signal would take about 25 000 years to reach our Sun from the Galactic Centre.

From our location, the other stars in the disc appear concentrated in a band across the sky: this is the Milky Way. These stars are orbiting around the Galactic Centre, taking more than 100 million years for a single circuit. Andromeda, our Galaxy's nearest major neighbour in space, is about 2 million light-years away. Like our own Galaxy, it contains stars of all ages, and gas.

A cube set down at random in our universe would need to have sides 10 million light-years long to contain, on average, one galaxy. The galaxies are not, however, sprinkled down randomly in space: most are in groups or clusters, pulled together by gravity. Our own Local Group, a few million light-years across, contains the Milky Way and Andromeda, together with at least 20 smaller galaxies. Gravity is pulling Andromeda towards us at about 100 kilometres per second. In about 5 billion years, these two disc galaxies will crash together.

Some clusters contain many hundreds of galaxies. Our Local Group lies near the edge of the Virgo Cluster, whose centre is 50 million light-years away. The clusters and groups are themselves arranged in still larger filamentary or sheetlike structures. One of the most conspicuous is the so-called 'Great Wall', a sheetlike array of galaxies about 200 million light-years away. Another huge concentration of mass, the 'Great Attractor', exerts a gravitational force that pulls us, along with the entire Virgo Cluster of galaxies, at several hundred kilometres per second.

A FRACTAL UNIVERSE?

Patterns as disparate as coastlines, mountain ranges, the branches of trees and the bronchial tubes in our lungs have a common feature – any small part, highly magnified, displays a pattern that is a simulacrum of the whole. The mathematician Benoit Mandelbrot coined the term 'fractal' to describe such patterns; through his insight we've come to appreciate the ubiquity of these structures in nature. So could our universe be a 'fractal', containing clusters of clusters of clusters . . . ad infinitum?

We now know that our universe actually isn't a 'fractal'. If hierarchical clustering actually continued indefinitely towards larger scales, then however deeply we probed into space, and however large a volume we sampled, the galaxies would still have a patchy distribution across the sky – we'd simply be sampling larger and larger scales in the hierarchy. But this is not how our universe looks. As we probe deeper into space, observing larger numbers of more distant galaxies, we see more clusters like Virgo, and more features like the 'Great Wall'. But we don't discover another 'layer' of still larger structures: the pattern becomes smoother. A box whose sides were 500 million light-years (dimensions still small compared with the observable universe) would be large enough to contain a 'fair sample' – wherever it was, it would contain roughly the same number of galaxies, grouped in a statistically similar way into clusters, filamentary structures and so forth. There is a genuine 'broad-brush' sense in which our observable universe – the region out to our horizon – is indeed homogeneous. There are many other 'Great Walls' further from us, but there don't seem to be any conspicuous features on still larger scales.

Ever since Copernicus, we've been reluctant to put ourselves 'centre stage' in the cosmos. If our location is typical, the large-scale universe (that is to say, everything more than a few hundred million light-years away) would look roughly the same viewed from any other galaxy.

THE EXPANDING UNIVERSE

Modern telescopes reveal galaxies as far away as 10 billion light-years. In this broad cosmological perspective, galaxies are just 'test particles' scattered through space indicating how the content of the universe is distributed, and how it is moving. When cosmologists speak of the 'expanding universe' their evidence comes mainly from the motions of galaxies.

This evidence dates back to the 1920s. Several astronomers made pioneering contributions, but the dominant figure was Edwin Hubble, after whom the Space Telescope has been named. He used the 100-inch telescope on Mount Wilson in California – the most powerful instrument of its time – to study the light from many galaxies. When this light is passed through a prism, it splits into a spectrum of different colours. Spectra of galaxies reveal patterns due to the distinctive colours of the light emitted or absorbed by the various kinds of atom (carbon, sodium and many more) that they are made of. Hubble found that the characteristic patterns were all stretched to somewhat longer wavelengths – shifted towards the red – compared with those measured in the laboratory, or in spectra from stars and gas in our own Galaxy. This redshift was larger for the fainter and more distant galaxies. Hubble claimed that the redshift of a galaxy was proportional to its distance. The best-known cause of a redshift is the familiar Doppler effect; if this is indeed the reason for Hubble's redshifts, then galaxies must recede from us with speeds proportional to their distance.

Theory had already foreshadowed the idea of universal expansion. In 1922, Alexander Friedmann, a Russian mathematician and meteorologist, showed that an expanding unbounded universe was consistent with Einstein's theory of general relativity. Einstein had originally resisted Friedmann's work – he favoured a static universe – but Hubble's discovery changed his attitude.[1]

Hubble's original evidence was actually far from convincing. The galaxies he studied were all relatively close by, so all he could directly infer was that our 'local supercluster' was expanding. It was a real leap of faith to believe that the 'Hubble law' applied out to distances hundreds of times greater, where the recession speeds would approach the speed of light. But, because Friedmann's work was already known, this interpretation was at least taken seriously right from the start.[2]

The expansion should *decelerate*, because of the gravitational attraction that each piece of matter exerts on everything else. If the density were sufficiently low, the deceleration would be gentle and the expansion would never stop. Our universe could then be infinite both in space and time. But a denser universe would eventually stop expanding, and collapse; although uniform (homogeneous) and unbounded, it could then be 'closed' and finite in both content and duration.

But in the 1930s – and for several decades thereafter – it remained unclear how well the large-scale universe could be described by these simple solutions of Einstein's equations; and there was no way to distinguish between the so-called 'closed' and 'open' versions.

Hubble's law now extends out to galaxies so far away that they are receding at more than 90 per cent of the speed of light. The simple 'model universes' turn out, more than 60 years later, to fit extraordinarily well – they are more relevant to our real universe than Friedmann and the other pioneers would have dared to hope.

We seem to be in an expanding universe, stretching out to a distance of 10 billion light-years, where distant objects get more and more widely separated as time goes on.[3] Observers in any other galaxy would witness a similar expansion away from them. There is nothing special about the location of our Galaxy in space. We are not, however, observing at a random *time*; the reasons for this will become clear later.

TIRED LIGHT?

The recession of distant galaxies obviously suggests that galaxies emerged from some kind of 'beginning', 10 or 20 billion years ago, when everything would have been crowded together. Hubble was, at least initially, surprisingly open-minded about whether his redshift law implied a real expansion. Could light traversing huge distances get 'tired', and reddened, even if the universe were really static? As late as the 1970s, some physicists in Paris were still seriously advocating '*photons fatigués*'; and the idea still resurfaces from time to time. It is therefore worth emphasising that 'tired light' has been discarded for specific reasons, rather than from a (perhaps prejudiced) reluctance to contemplate some fundamentally new effect.

Light is, physically, a wave of electrical and magnetic energy travelling through space. The wavelengths depend on the colour, being shorter for blue and longer for yellow and red light. The wavelengths of all types of light from a distant galaxy are increased by the same factor. This is exactly what would be expected in an expanding universe – wavelengths, as in the normal Doppler effect, are all stretched in the same proportions. But a mechanism for 'tiring' the light would generally not do this. Moreover, any effect whereby light lost energy by repeatedly scattering off hypothetical particles would blur the images of distant objects, also contrary to observation.

There is now a further test. A clock receding from us appears to run slow – if it sends out periodic 'beeps', the later ones have further to travel, and so the intervals between their arrival are lengthened. This slowdown is directly related to the redshift. The successive 'wavecrests' in the light from any atom or molecule are due to its vibrations, which are essentially a microscopic clock – the wavecrests arrive more slowly when the source is receding, and the wavelengths are stretched. Nature provides us with clocks bright enough to be seen in distant galaxies – supernovae. One particular type of supernova (unimaginatively called 'Type 1'), signalling the death of stars in violent thermonuclear explosions, brightens and fades in a standard way. Remote 'Type 1' supernovae do indeed appear to flare up and fade more slowly than nearby ones. The slowdown is just what we would expect if they are receding, and reddened by the Doppler effect, but has no natural explanation in a static universe.

A static universe would actually entail even worse paradoxes than any big-bang theory. Stars don't have infinite energy reserves: they evolve, and eventually exhaust their fuel. So, therefore, do galaxies, which are essentially aggregates of stars. The inferred ages of our Milky Way and other galaxies are about 10 billion years – entirely consistent with the view that our universe has been expanding for only about that long. If our universe were static, all galaxies must have mysteriously 'switched on', in their present positions, in a synchronised fashion about 10 billion years ago. A non-expanding universe – even the kind that Einstein espoused before he learned about Hubble's work (further discussed in chapter 8) – entails severe conceptual difficulties.[4]

AN EVOLVING UNIVERSE?

Astronomers can study parts of space whose light set out a long time ago. If we lived in a wildly irregular universe, these remote regions might bear no resemblance to our own locality. But since our universe (or at least the part we can see) has a 'broad-brush' uniformity, resembling a seascape rather than a mountain landscape, we infer that all parts have evolved the same way and had similar histories. So that when we observe a region lying (say) 3 billion light-years away, its gross features (what galaxies look like, the way they are clustered and so on) resemble our own locality as it would have looked 3 billion years ago.

Were the galaxies more closely packed together in the past? And do remote galaxies look different, as we'd expect if they were, on average, younger when they emitted the light now reaching us? Edwin Hubble's early data couldn't provide an answer. These questions are important because the answer need not be 'yes' – an expanding universe need not necessarily evolve. This point was forcefully made by Fred Hoyle, together with Hermann Bondi and Thomas Gold, two theorists who had come to Cambridge as refugees from Austria. Bondi, primarily an applied mathematician, contributed influential ideas to astronomy and the theory of relativity. Gold's range of expertise was more eclectic. His academic career was launched by a thesis on hearing and the physiology of the inner ear; he went on to deploy his physical insights in many areas (including neutron stars, as described in Chapter 4).

Bondi, Gold and Hoyle conjectured that we might live in a 'steady-state' universe, in which continuous creation of new matter and new galaxies maintained an unchanging cosmic scene despite the overall expansion. Individual galaxies would still evolve; but as they aged they would disperse more widely, and new younger ones would form in the gaps that opened up between them. The universe, having had an infinite past, might have achieved some unique self-sustaining state. The required creation rate was so low that it would have been entirely undetectable – one atom per century in each cubic kilometre – but many found the concept 'ad hoc' and implausible. Hoyle countered this objection by developing a specific theory to describe how new atoms could occasionally 'materialise'; in any case, he argued, the creation of everything 'in one go' was an even greater leap beyond conventional physics. (Bondi, Gold and Hoyle came up with the steady-state idea in 1948, after seeing a film called *The Dead of Night*, whose conclusion recapitulated the opening scene.)

The steady-state theory provided a constructive stimulus for more than 15 years. If a steady state prevailed, then distant regions, even though we see them as they were a long time ago, should, statistically, look *just like* nearby regions – this is a very specific prediction. If remote galaxies look different on average, we can't be living in a steady-state universe.

THE LOSING BATTLE FOR A STEADY STATE

Even if our universe is evolving, changes would be so slow that they would only be manifest over billions of years. To detect an evolutionary trend (or to check whether the universe is really in a steady state) one must probe galaxies so far away that their light set out several billion years ago. Such efforts started as early as the 1950s, using the telescope on Mount Palomar in California (which, with its 200-inch-diameter mirror, was then much the largest in the world). The results were inconclusive. Normal galaxies with sufficiently high redshift were not luminous enough to register on photographic plates, even with such a powerful light-gatherer as the 200-inch telescope.

The world's best optical telescopes, in the 1950s, were concentrated in the US, particularly in California. This shift from Europe had come about for climatic as well as financial reasons: it plainly made little sense to build elaborate telescopes on low-lying sites exposed to the British climate. However, the next observational breakthrough in cosmology (after Hubble's discovery of the cosmic expansion) came from a quite different technique – radio astronomy. Radio waves from space can pass through clouds, so Europe had no climatic handicap in this new science.

Even in the early 1950s, when techniques were still primitive, radio astronomers in Britain and Australia had discovered a specially intense 'hiss' when their antennae were pointed in particular directions. Some of these cosmic radio sources could be readily identified. For instance, strong radiation emanated from the Galactic Centre; another strong source was the Crab Nebula, the expanding debris of a supernova explosion witnessed by oriental astronomers in AD 1054 (described in Chapter 1).

In 1954 two Californian astronomers, Walter Baade and Rudolf Minkowski, discovered that the second-strongest radio source in the sky was an unusual remote galaxy. Its radio emission was so intense that a similar galaxy could have been picked up by radio telescopes even if it were several times further away; but its visible light would then have been too faint to register. Baade and Minkowski's discovery meant that the new techniques of radio astronomy offered a deep probe into the early universe: radio telescopes could pick up emission from some unusual 'active' galaxies (now believed – see Chapter 5 – to harbour massive black holes in their centres) even when they were too far away to be seen with optical telescopes.

Radio telescopes are amazingly sensitive to very weak signals. The pioneer radio astronomer Martin Ryle had a nice way of illustrating this. When 'open days' were held at his observatory just outside Cambridge, each visitor was asked to take a tiny slip of paper from a pile. On it was written 'in picking this up you have expended more energy than has been received by all the world's radio telescopes since they were built'.

A problem in the early days of radio astronomy was to pin down the exact directions that cosmic 'radio noise' came from. Ryle invented a technique that ameliorated this problem, enabling him to survey the northern sky and locate several hundred sources. He used his data, very ingeniously, to conclude that our universe was actually evolving, and couldn't be in a steady state.

Ryle didn't know the distances of his radio sources (most had no visible counterpart, so optical astronomers couldn't measure redshifts), but he assumed that the weaker sources were, on average, further away than those giving intenser signals. He counted the numbers with various apparent intensities, and found that there were surprisingly many apparently weak ones – in other words, sources mainly at large distances – compared with the number of stronger and closer ones. It seemed as though we were in the middle of a huge sphere, several billion light years in radius, and there was a much higher concentration of radio sources near the surface of the sphere than near its centre. This seemed incompatible with a steady-state universe, where the sources must, by hypothesis, belong to similar populations at all times, and therefore at all distances. However, it was quite compatible with an evolving universe. Ryle conjectured that galaxies were more prone to undergo the mysterious outbursts that generated intense radio waves when they were young, several billion years ago. If galaxies have now matured and 'quietened down', fewer nearby ones would be detected as radio sources.

Ryle's argument was first put forward in the 1950s, and provoked a noisy (and often ill-tempered) controversy that ran for several years. When I became aware of it in the early 1960s, Ryle's case seemed ingenious and compelling, and the persistent opposition of the 'steady statesmen' perplexed me. Only later did I learn the background context. In the early 1950s, Ryle had been equally dogmatic about other issues, on which he had been on shaky ground.

For instance, when radio sources were first discovered, Ryle thought they were 'radio stars' within our own Galaxy. They didn't seem concentrated towards the plane of the Milky Way; but this could have meant that they were *very* close (by astronomical standards): if the detectable sources were all closer than the *thickness* of the galactic disc, only a few hundred light-years, they would appear uniform over the sky. Gold and others contended that the sources weren't concentrated towards the Milky Way because they were nothing to do with our Galaxy, and much further away. Ryle initially resisted this proposal with great fervour (even though the vast distance of these objects later became pivotal to his cosmological arguments).

A second reason for scepticism was that Ryle's earliest radio surveys had turned out to be flawed – they yielded such a blurred map of the radio sky that two or more separate sources were sometimes counted as one. However, by 1958, when Ryle presented his case for an evolving universe in a major lecture at the Royal Society, the most serious 'bugs' had been dealt with, and his data were reliable; essentially everything he said in that lecture has stood the test of time.

The steady-state theory called some cherished beliefs into question, and offered specific predictions which goaded the observers into attempts to refute it. The theory's originators, an articulate and inventive trio who relished controversy, were effective publicists. Hoyle, in particular, was a brilliant populariser; many younger cosmologists (and I'm among them) owe their original impetus to his books and broadcasts. The confrontation between a steady-state universe and an evolving one accordingly achieved wide public currency. At least this was true in Britain: the voices of Bondi, Gold and Hoyle failed to carry across the Atlantic, and their theory was never taken very seriously in the US. But it was the pioneer radio astronomers in Britain and Australia (many of whom acquired their expertise working on radar during the Second World War) who were best placed to carry out the crucial radio surveys.

Ryle plainly wanted his radio surveys to have a decisive impact on cosmology, which they could do only by refuting the 'steady state'. He had invested years of effort in designing and building new instruments, as well as in the data-gathering itself. No single individual nowadays can master all the necessary techniques; Ryle was an exceptional exemplar of the pioneering radio astronomers who conceived and built novel equipment, and themselves drew fundamental inferences from the data. Nobody would persevere with such a challenging project without (perhaps excessive) confidence in its potential importance or decisiveness. Demanding and sustained research programmes tend to be driven by just such personalities.

Despite Ryle's compelling arguments in 1958 (or so they seem in retrospect), controversy took several more years to die down. The radio sources remained an enigma. They were thought to be a special kind of galaxy, but only a few relatively nearby 'radio galaxies' had actually been seen by optical astronomers. Ryle argued that the rest were similar, but lay beyond the range of optical telescopes. There was no corroboration (by, for instance, redshift measurements, or association with remote galaxies) for their alleged immense distances. And nobody yet had any idea how a galaxy could channel such immense power into radio waves.

So could these mysterious sources be, after all, a nearby population within the Milky Way, as Ryle himself had believed until 1954? The counts would then be probing the 'geography' of our own Galaxy, and irrelevant to cosmology. The main proponent of this 'escape route' was Dennis Sciama, who became (to my great good fortune) my research supervisor when I started as a graduate student in 1964. He then described himself (probably correctly) as the only remaining believer in the steady-state theory apart from the trio who had invented it. But even Sciama capitulated, when confronted with new evidence from optical astronomy. Some of Ryle's sources turned out, after all, to have visible counterparts. They were 'quasars' – galaxies whose 100 billion stars are outshone by concentrated radiation in their centres (which is now generally believed to involve a huge black hole). These quasars had very large redshifts. Sciama accepted that high redshifts indicated immense distances, and quasars proved to be more common at high redshifts than nearby, corroborating Ryle's conjectures. Even 30 years later we're still trying to understand *why* the quasar population evolves in this way.

The radio source counts are now of historical interest, having long ago been superseded by more informative and less ambiguous techniques. They were, nevertheless, the first real 'cosmological test'. The episode gave me an early opportunity, while I was a Cambridge student, to follow a scientific controversy at close range, and to appreciate the contrasting attitudes that cosmological debates generally manifest. Sciama was committed to the steady-state theory. For him, as for its inventors, it had a deep philosophical appeal – the universe existed, from everlasting to everlasting, in a uniquely self-consistent state. When conflicting evidence emerged, Sciama therefore sought a loophole (even an unlikely-seeming one), rather as a defence lawyer clutches at any argument to rebut the prosecution case.

When a novel phenomenon is discovered, and data are still sparse, many different theories may be in the running to account for it. Closer scrutiny of the theories may suggest new ways of testing some of them; it may reveal internal inconsistencies in others. Such efforts narrow down the range of tenable explanations and perhaps identify a 'front runner'. Some theorists, like Sciama, are enthusiastic backers of a particular hypothesis – such a commitment is, for them, a necessary motivation. Others hedge their bets, and may indeed investigate two or more different (and mutually inconsistent) hypotheses in parallel – for them, the quest for deeper understanding is sufficient motivation. Hoyle, the most creative and original astrophysicist of his generation, was certainly in the latter category. He favoured a steady-state universe, but this didn't inhibit him from contributing key ideas to rival theories.

QUASARS, THEN AND NOW

In the early days of radio astronomy, it was hard to locate the sources accurately enough to be sure which visible object was generating the radio emission (unless there was some bright and obviously peculiar object in the same direction, like the Crab Nebula). A major advance came in 1963, when Cyril Hazard and his colleagues, working in Australia, managed to pin down the position of one particular source to a precision of a few arc seconds. They pointed the Parkes radio telescope towards 3C273 (so called because it was number 273 in Ryle's third Cambridge catalogue of sources) while the Moon moved across that part of the sky. The source 'disappeared' when the Moon's limb passed in front of it; they noted the exact time (and the exact position of the Moon) when this happened. The only visible object close to the right position was a rather faint (13th magnitude) starlike object.

Cyril Hazard didn't himself have access to a large optical telescope, so he urged Maarten Schmidt, a Dutch astronomer working in California, to take the optical spectrum of 3C273 with the Mount Palomar 200-inch telescope. Schmidt found strong features in the spectrum, which he could not immediately understand – the object was certainly not a star. He was mystified until it 'clicked' that the spectrum had the pattern expected from glowing hydrogen, except that the features were all shifted to wavelengths 15 per cent longer than they should have been. A shift like this had hitherto been seen only in very remote (and faint-seeming) galaxies. But 3C273, which appeared point-like and so was certainly smaller than a galaxy, was much *brighter* than an *entire* galaxy with such a substantial redshift should have been.

Once the intellectual block had been removed, several other hitherto mysterious spectra of similar sources began to make sense. A new class of objects had been discovered that looked like ordinary stars on photographic plates, but whose spectra showed high redshifts. They came to be known as quasi-stellar objects, or 'quasars'. Knowing the redshifts, one could infer the distances from Hubble's law. To appear so bright at such great distances, quasars would have to outshine entire galaxies even though they are energized by something smaller than our Solar System. They undergo outbursts, which are equivalent to turning on 10 000 galaxies like our own within a single day. How this happens is still poorly understood.

NEW PHYSICAL LAWS?

When quasars were discovered they certainly posed a challenge. It took several years to realise that they lay in the centres of galaxies, and were related to the radio galaxies whose remarkable properties Ryle had already recognised. In the early days, some researchers doubted whether quasars could be accommodated within the framework of conventional physics. Maybe some 'new physics' was operating: either a different explanation of redshifts so that quasars could be nearer (and therefore not so excessively powerful), or else a more efficient energy-production mechanism. This was then an entirely realistic concern, and it triggered a lively debate.

It is in principle conceivable that these cosmic phenomena involve some basically new law of nature. After all, a physicist whose laboratory floated freely in space would probably never have discovered gravity, because this force is very weak unless a large mass such as the Earth is involved. Maybe, therefore, there are other effects, insignificant even on the scale of the Solar System, which are nonetheless crucial in galactic centres or cosmology.

Various puzzling observations were adduced in support of such a case. For instance, quasars were claimed to be close in the sky to low-redshift galaxies too often for this to be a chance superposition, implying a physical link. But it is all too easy to perceive patterns in random data – patterns that may seem improbable. The crucial test is whether a hypothesis has predictive power, and applies not just to the objects in which the alleged effect was first noticed, but to some new samples as well. Most of the apparent anomalies were diluted (and sometimes even disappeared) as more data accumulated.

New peculiarities have often surfaced to replace the old. As studies continue, *more and more* surprising effects are bound to be discovered. Unless these effects can nearly all be incorporated into a single theory, they don't add cumulative weight to an unorthodox viewpoint. Moreover, when there is said to be only one chance in (say) 200 of getting a particular correlation by pure coincidence, we should not blindly accept this as significant without applying a 'discount' to allow for all the other similar effects, which might equally well have been found but were not – claims for extra-sensory perception (telepathy and the like) commonly arise from overlooking such points.

A lively conference addressing these issues took place in 1970; it was held (somewhat incongruously) at the Vatican, under the sponsorship of the Pontifical Academy of Sciences. Some participants displayed pictures of galaxies with very different redshifts, claiming that these were physically linked, rather than being just chance superpositions of foreground and background objects, and that the redshifts were therefore 'anomalous'. 'I don't see the logic of rejecting data because they seem incredible,' commented Fred Hoyle. 'I can see no *better* reason' was the response of Lyman Spitzer, an eminent theorist from Princeton (and a more conventional one).

The debate about whether quasars are 'near or far' was reminiscent of an astronomical controversy that took place two hundred years earlier concerning the reality of binary stars. Many instances were known where pairs of stars lay close together on the sky, and John Michell (who will feature again in Chapter 5) showed statistically that there were *too many* such pairs for them to the merely chance superpositions of foreground and background stars. He therefore argued that these stars must be physically associated 'either by gravity . . . or by some other law or appointment of the creator'.

William Herschel disagreed. He believed (wrongly, as we now recognise) that all stars were equally luminous. Since the stars in each alleged binary were generally unequal in brightness, he therefore concluded that one must be much further away than the other, so they could not be orbiting around each other. Herschel took 36 years to change his mind. There is a parallel between this debate and the controversy in the late 1960s between those who believed that the redshifts of quasars were a true measure of their distance, and those who adduced statistical evidence to the contrary.

This debate about quasars highlighted the contrasting attitudes of different scientific personalities. Many would have been genuinely disturbed if anomalous redshifts really existed, because it would have meant that we were further from a definitive picture of our universe. On the other hand, those who espoused radical views would have been elated if astronomical observations had revealed some fundamentally new physics. Philosophers of science would be surprised at how many astronomers are eager rather than reluctant to join a revolutionary bandwagon.

My own attitude, apparently not widely shared, was that of a *reluctant* conservative. I wished the radicals had been right, but was sceptical about their arguments, and doubtful that the need for new physics had been justified. Many aspects of quasars were problematical (some, indeed, still are today). But the same can be said of many things that have been studied far more intensively, and for much longer. It is still unclear why the number of sunspots goes up and down on a roughly 12-year cycle; and laboratory phenomena such as superconductivity still flummox physicists. Nobody seriously invokes 'new physics' in any of these cases. Astrophysics is a field with a high ratio of problems to practitioners, and it would be astonishing if there were not still many mysteries. Even when progress has seemed slow, it has never signalled such an impasse as to justify abandonment of 'conventional' physics, whose rich consequences are far from having been exhausted.

Only in the earliest stages of the big bang, and deep inside black holes, are we confronted with physics that is 'new', in the sense that it can be checked neither experimentally nor even by observing 'extreme' astronomical objects like quasars or supernovae.

WHEN GALAXIES WERE YOUNG

Cosmologists depend on observations rather than experiments. They resemble paleontologists or geologists, trying to infer how our Earth, and the creatures on it, have evolved. Cosmologists study 'fossils' of the past (old stars, chemical elements synthesised when our Galaxy was young, and so forth). But they actually have an edge over practitioners of the other 'historical' sciences: by directing their telescopes towards distant objects, they can complement the 'fossil' evidence by actually *observing* the past.

Some critics, those of a creationist persuasion, deride Darwinism as 'just a theory'. What they mean is that it is founded on indirect inferences – indeed it is, though those inferences interlock into an exceedingly robust case. But cosmologists can actually *see* the evolution they claim – distant galaxies, whose light set out several billion years ago, manifestly look different from their counterparts nearby. This is not, of course, a 'time machine' of the kind that leads to the paradoxes ('killing your grandmother in her cradle' and suchlike) further discussed in Chapter 13. We are not probing the history of our own locality; but we are seeing snapshots of many distant galaxies which should, statistically at least, look similar to how our Milky Way, the Andromeda Galaxy and other nearby systems would have looked billions of years ago.

These galaxies can be clearly imaged by the Hubble Space Telescope. Competition for access to this instrument is so keen that even those who succeed are generally granted only a few hours of observing time. However, the director of the project, Robert Williams, is allocated, ex officio, a quota to use at his own discretion. He seized this enviable opportunity, and decided to point the telescope for ten whole days towards the same small piece of sky. This long exposure has yielded the sharpest and most detailed images of the distant universe that we have yet seen. Close-packed over the sky are faint galaxies, each seeming so small that it would be a barely perceptible blurred smudge on a picture taken from the ground. These objects, with a wide variety of shapes, are a billion times fainter than any star we can see with the unaided eye. But each is an entire galaxy, tens of thousands of light-years in size, which appears so small and faint because of its huge distance.

What is fascinating about these pictures is not the record-breaking distance in itself, but the *huge span in time* that separates us from these remote galaxies. They look different from their nearby counterparts because they are being viewed when they have only recently formed: they have not yet settled down into the steadily spinning 'pinwheels', like the beautiful nearby spiral galaxies depicted in most astronomy books. Some consist mainly of glowing diffuse gas, which hasn't yet fragmented into individual 'droplets', each destined to become a star. They appear intrinsically bluer than present-day galaxies (after correcting, of course, for the redshift), because massive blue stars, which would all by now have died, were still shining, transmuting pristine hydrogen into the elements of the periodic table, when the light left these distant galaxies.

The Space Telescope is showing us what our Milky Way would have been like when its first stars were shining brightly. There would then have been no complex chemistry, no planets and (presumably) no life. But these marvellous images offer new insight into the remote era when the basic building blocks of our Solar System were being laid down.

THE QUASAR ERA

The most distant quasars are so far away that the redshift has stretched the wavelengths of their light by as much as six.[5] Hardly any of the galaxies imaged by the Space Telescope are as far away as this. The ultra-remote quasars show up so brightly only because of their enormous intrinsic brightness. How long ago did that ancient light leave the quasars? If our universe had expanded at a steady speed – with neither acceleration nor deceleration – then when it was a sixth of its present scale (in the sense that distances were scaled down six times smaller) it would have been a sixth of its present age.[6] However, the expansion has probably been decelerating, because of the gravitational pull that everything in the universe exerts on everything else. The earlier stages of cosmic expansion would consequently have been, relatively, even shorter. So, when observing remote quasars, we are looking back to when the universe was probably less than a tenth, rather than about a sixth, of its present age. Astronomers can therefore probe the last 90 per cent of cosmic history. The existence of these quasars tells us that, by the time the universe was about a billion years old, some galaxies (or at least their inner regions) had already formed.

Quasars are probes of the remote past. They have turned out to be of even more fundamental interest because at their cores lurk vast black holes – places where space itself has been 'punctured' by some colossal collapse, and whose interiors may hold the secrets of how our universe began, and even of its links with other universes.

But what about still earlier epochs, before any galaxies could have formed? How, for instance, did the dominant present-day structures in our universe – galaxies and clusters – emerge from amorphous beginnings in the early universe? Did everything really emerge from a dense 'beginning' 10 or 20 billion years ago?

Georges Lemaître, a Belgian priest who became president of the

Pontifical Academy of Sciences, speculated about this soon after the expansion of the universe was discovered. He presented his 'primeval atom' scenario in these picturesque terms:

> The evolution of the universe can be likened to a display of fireworks that has just ended: some few wisps, ashes and smoke. Standing on a well-chilled cinder, we see the fading of the suns, and try to recall the vanished brilliance of the origin of the worlds.

The traces of that vanished brilliance were revealed in 1965.

NOTES

1 Einstein's belief in a static universe had led him to introduce an extra feature in his equations – a 'cosmic repulsion' force which could counterbalance gravity on the cosmic scale. This concept has recently been resuscitated, as described in Chapter 8.

2 The region that Hubble actually studied is small enough that Newtonian theory describes its dynamics quite adequately – indeed, the error in so doing is only one part in 100 000. The population of galaxies stretches out to greater distances; we can imagine a series of spherical shells around us, the larger ones receding faster. The recession of the larger shells may approach the speed of light, and a Newtonian calculation is no longer adequate. Friedmann's work took these effects into account, and offers an adequate description provided the universe is sufficiently uniform in its density and expansion rate. Local regions can still, quite validly, be envisaged in Newtonian terms: a sphere 'hollowed out' from a Friedmann universe obeys Newton's laws exactly. Almost all the calculations of how galaxies form, and how they move in clusters and superclusters, use just the familiar inverse square law of Newtonian gravity (see Chapter 7).

3 There is no expansion within individual galaxies, nor even within clusters of galaxies. The simple 'Hubble law' applies only on scales large enough for the universe to be treated as smooth. In practice, there are significant deviations from the simple law on scales up to those of superclusters.

4 When quasars were discovered in the 1960s, there was renewed lively debate about whether the redshifts of these unusual objects might have some special origin. It took a few years before the evidence convinced most astronomers that quasars also manifest Hubble redshifts.

5 The light we detect from these quasars was actually emitted in the far ultraviolet. Such radiation cannot be detected by the eye, nor can it even penetrate the Earth's atmosphere. But the ultraviolet radiation from these exceedingly remote objects, by the time it reaches us, has been shifted into red light.

6 This kind of assertion at first sight seems perplexing if the speed of light

is an ultimate 'speed limit': doesn't it imply that the quasar moved away at five times the speed of light, if the light has taken five-sixths of our universe's present age to get back to us? Einstein's special theory of relativity tells us that nothing can move faster than light, when time is measured by a clock that *isn't* sharing in that motion. But that theory also tells us that a fast-moving clock runs slow. (This is the basis of the 'twin paradox' described in Chapter 13.) A fast clock could indeed travel five light-hours for every hour that it records: it would need to go at about 98 per cent of the speed of light.

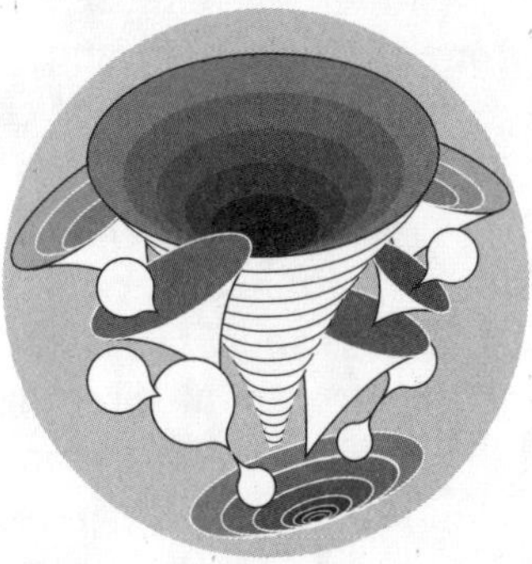

3

PREGALACTIC HISTORY: THE CLINCHING EVIDENCE

This is *our* Universe, our museum of wonder and beauty, our cathedral.

J.A. Wheeler

THE AFTERGLOW OF THE COSMIC FIREBALL

In a 1970-vintage textbook by the Canadian cosmologist James Peebles, one chapter is entitled 'Golden moments in cosmology'. There were then only two of these. The first was Hubble's realisation that our universe was expanding (as described in the previous chapter). The other was the detection by Arno Penzias and Robert Wilson of the 'afterglow of creation' – prosaically called the cosmic microwave background radiation. This discovery, made at the Bell Telephone Laboratories in 1964–65, with a sensitive antenna intended for use with the Echo communication satellites, was accidental: indeed Penzias and Wilson did not immediately appreciate what they had found. They had discovered that intergalactic space wasn't completely cold. Their antenna was picking up microwaves – electromagnetic waves of the kind that generate heat in a microwave oven, but very much less intense – which seemed to arrive with equal strength from all directions and had no obvious source: not the atmosphere, nor known types of cosmic radio source. The warmth is a relic of Lemaître's 'vanished brilliance' – of a dense 'fireball' phase when everything was hot, dense and opaque.

Penzias and Wilson's paper announcing 'excess antenna temperature at 4080 Mc/s' appeared in the 1965 *Astrophysical Journal.* The excess amounted to about three degrees above absolute zero. This is of course exceedingly cool (minus 270 degrees Celsius); but there's a well-defined sense in which intergalactic space still contains a lot of 'heat'. Every cubic metre contains about 400 million quanta of radiation, or photons: in comparison, the average density of *atoms* in the universe is only about 0.1 per cubic metre. This latter number is less precisely known, because most atoms may be in diffuse gas or 'dark' matter, but there seem to be at least a billion photons for every atom in the universe.

The only plausible explanation for the microwave background is that it has survived from an epoch when our entire universe was hot, dense and opaque. Its discovery quickly led to a general acceptance of this radical concept.

(This realignment of cosmological attitudes has some parallels with a sudden and drastic shift of opinion among geophysicists at about the same time. Continental drift had long been advocated by Alfred Wegener – who was, like the pioneer cosmologist Alexander Friedmann, originally a meteorologist. But the evidence before 1965 was no more than suggestive; geophysicists had no plausible explanation for how it might come about, and were mostly dismissive of the idea. A favourable consensus suddenly emerged after the Cambridge geophysicists Drummond Matthews and Frederick Vine showed convincingly that the sea floor was spreading out from the mid-Atlantic ridge, pushing Europe and America apart.)

Penzias and Wilson's discovery was preceded by a chapter of accidents, misunderstandings and communication failures. George Gamow, a Russian émigré in the US, was, along with Georges Lemaître, a leading early advocate of what is now called the big bang. In the 1940s Gamow, with his students Ralph Alpher and Robert Hermann, had proposed that the universe started off very hot – they even calculated that the present temperature might be about five degrees above absolute zero – but they did not initiate any experimental search for the 'relic' radiation, even though it might have been feasible even at that time.[1]

In the 1950s radio astronomers, in both France and Russia, had noted a background 'hiss' that couldn't be attributed to instrumental effects or known sources. In 1961, there was another experiment, this time by Edward Ohm in America. Yakov Zeldovich and other Russian cosmologists read about this; but they misunderstood what Ohm, a radio engineer, meant by 'sky background', and wrongly inferred that he was *ruling out* a temperature of more than one degree.

Igor Novikov and Andrei Doroshkevich in Moscow realised, in 1962, that the 'relic' radiation from the big bang might be detectable: they even noted that the antenna at Holmdel, New Jersey – the one used by Penzias and Wilson – was well suited to this experiment. Novikov now heads his own institute in Copenhagen, and continues as a leader of cosmological research. He thinks that earlier theorists were unduly pessimistic about the chances of detecting this radiation because they had realised, correctly, that its energy was no higher than that carried by starlight or by cosmic rays (fast-moving particles, mainly produced by supernova explosions, which pervade the whole Galaxy). Since the latter energies were themselves uncertain, it seemed unfeasible, at first sight, to isolate a 'cosmological' background. But this overlooks the point that the primeval radiation, unlike the other backgrounds, would now be concentrated in the 'microwave' band (wavelengths of centimetres or millimetres), and would have a distinctive spectrum.

The first people to search systematically for 'cosmic background' radiation were Robert Dicke and his colleagues at Princeton. The idea of an *oscillating* universe appealed to Dicke. He worried, however, about the cumulative build-up of 'nuclear waste' on a cosmic scale. Stars derive their power from converting hydrogen into helium, and then processing the helium further up the periodic table. If there had already been many cycles, why wouldn't everything by now have been transmuted into iron? Dicke tried to evade this objection by arguing that, in the bounce, everything got so hot, and the collisions among the particles were so violent, that iron nuclei (each made of 26 protons and 30 neutrons) would be broken up; each cycle would then start with a 'new deal' of hydrogen and helium.

Dicke's theories have now been superseded: if a new universe emerges phoenix-like from the collapse of the old, the physics of the 'bounce' is so exotic that no individual particles (and maybe no 'memory') from the previous cycle survive at all. But these ideas motivated him to search for relics of a hot 'primordial fireball'. Though deeply interested in cosmology and gravitation, Dicke was primarily an experimentalist, and had just the technical expertise to build the 'radiometer' that was needed.

Penzias and Wilson had little prior interest in cosmological theory, so there is no reason why they should have been aware of Alpher, Hermann and Gamow's researches; but the oversight is much more surprising in Dicke's case. (Novikov's work, published in the Soviet literature, could understandably have been overlooked, but there seems less excuse when the papers appeared in premier American scholarly journals, as those of Gamow's group did.)

Had it not been for Dicke, Penzias and Wilson might not have realised what they had found. When Dicke heard about the results from the Bell Laboratories, his reaction was, 'We've been scooped'. And very soon afterwards the Princeton group made a measurement of their own, corroborating the discovery. This whole episode demonstrated the fitful and unpredictable way in which breakthroughs occur. The discovery gladdened Lemaître, who was told about it just weeks before he died. The one cause for regret is that Dicke, an outstanding intellect, unique in combining theoretical background with just the right experimental expertise, was a secondary corroborator rather than (as would have been fitting) the prime discoverer of the background radiation.

Dicke and his Princeton colleagues spelled out what it all meant in the same issue of *Astrophysical Journal* as Penzias and Wilson announced their discovery. According to this now standard story, everything in our universe – all the stuff that galaxies are now made of – was once an exceedingly compressed and hot gas (hotter than the Sun's core). The intense radiation in this fireball, though cooled and diluted by the expansion, would still be around, pervading the whole universe: the microwaves are an echo, as it were, of the 'explosion' that initiated the universal expansion.

Robert Wilson later noted that he himself only fully realised the import of what he had discovered when he read a 'popularised' account in the *New York Times*. This reaction highlights an occupational risk facing all scientists. Generally, researchers don't shoot directly for a grand goal. Unless they are geniuses (or cranks) they focus on bite-sized problems that seem timely or tractable. That's the methodology that pays off for most of us. But we may then forget that we are wearing blinkers, and that our piecemeal efforts are worthwhile only insofar as they're steps towards some fundamental question. The response of non-specialists helps us to see our work in perspective.

STEPS TO COBE

In its early 'fireball' stage our universe would have been an almost featureless gas: atoms, mixed with radiation quanta (photons). At the high prevailing temperatures, atoms would be torn apart into their constituent electrons and nuclei. The photons, repeatedly absorbed and scattered, would have come into equilibrium with their surroundings: they would have turned into what physicists call 'black body' or 'thermal' radiation. Once its intensity has been measured at one wavelength, a straightforward formula tells us how intense such radiation should be at any other wavelength. In the years after Penzias and Wilson's discovery, at least 30 separate measurements were made at different wavelengths; all seemed consistent with thermal radiation. Most of these measurements were made from the ground at centimetre wavelengths. However, if the radiation is indeed thermal, its energy is concentrated at wavelengths around two millimetres. The most crucial measurements, therefore, are in the millimetre rather than centimetre band. Unfortunately, the Earth's atmosphere is opaque at these shorter wavelengths, and right until 1990 the direct measurements, made using balloons and rockets, were ambiguous and uncertain.

The Cosmic Microwave Background Explorer (COBE), a satellite carrying experiments specifically designed to make precise millimetre-wavelength measurements, ended this uncertainty. John Mather and his colleagues found that the spectrum follows a black body with an accuracy better than one part in 10 000. The temperature is 2.726 K. When the first COBE results were announced at the American Astronomical Society in 1990, the

1500-strong audience greeted this magnificient achievement with prolonged applause. There could thereafter be no reasonable doubt that this radiation was indeed a relic of the dense fireball from which our expanding universe emerged.

Think of a box whose sides, of length *R*, expand with the universe. Insofar as the universe is homogeneous, the contents of this box – matter (electrons and nuclei) as well as photons – would be a representative sample. The wavelengths of any radiation in the box stretch in proportion to *R*, and the temperature goes as $1/R$. It would have taken about half a million years for expansion to cool the primordial mixture to 3000 degrees – somewhat cooler than the Sun's surface. The electrons would by then have slowed down enough to attach themselves to nuclei, forming neutral atoms, which cannot scatter the radiation as efficiently as free electrons could during the earlier and hotter stages. The primordial material would thereafter have been transparent. The microwave photons reaching COBE and other instruments are direct messengers from the era when our universe was more than a thousand times more compressed – at 3000 degrees rather than 2.7 degrees, and long before any galaxies came into existence. But the photons, now diluted, and shifted into the microwave part of the spectrum, are still around: they fill our universe and have nowhere else to go.

THE HELIUM PROBLEM: CORROBORATING A 'HOT BIG BANG'

The microwave background is a 'fossil' from very early stages in the expansion. There seems to be another important vestige of the primordial fireball in the present-day universe: the element helium, which comprises about a quarter of the mass of most stars, including the Sun.

Until the four authors known as B^2FH (see Chapter 1) proposed how atoms were transmuted inside stars, there was no serious explanation for the regular proportions of the various chemical elements in our Solar System and in other stars. Lemaître and Gamow had conjectured that the whole mixture was 'cooked' in the initial instants of cosmic expansion. Not enough nuclear physics was then known to quantify this idea. It turns out not to work: the expansion would be too fast to allow time for the requisite series of reactions. Moreover, if the chemical elements were all primordial (in the sense that they predated stars and galaxies) the abundances would be the same everywhere: it would be a mystery why the oldest stars contain a smaller proportion of heavy elements relative to hydrogen – something that is, of course, very natural if the elements accumulated gradually over galactic history.

Fred Hoyle (the H of B^2FH) actually preferred a steady-state universe. But it was Hoyle, in a celebrated series of radio talks, who popularised the term 'big bang' for the rival theory. This flippant description expressed his distaste for Lemaître and Gamow's idea. It was no accident that the key idea that the atoms of the periodic table resulted from transmutations in stars came from a 'steady statesman': the central dogma of the steady-state theory was that every strand of cosmic evolution must be going on, somewhere, now. The sites for element formation had to be sought in the present universe: they couldn't be relegated to an unobservable early era (the big bang). The steady-state theory was generally abandoned by the late 1960s, for reasons described in Chapter 2. But it has left an enduring legacy insofar as it stimulated the successful concept of nucleosynthesis in stars.

In one respect, though, Lemaître and Gamow have been vindicated. Stellar processes cannot account for *all* the elements. Helium, in particular, poses a puzzle. At first sight, one might expect helium to be the easiest element to explain, because stars spend most of their lives fusing hydrogen into helium. However, most helium made in stars is processed further up the periodic table before being recycled back into interstellar space (and thence into new stars). The 'heavy' elements, all together, amount to only one or two per cent of the hydrogen in the Solar System and in similar stars.[2]

If our galaxy had started off as pure hydrogen, then only a few per cent would have been 'processed' by nuclear reactions before the Solar System was born; there would be a still lower proportion in old stars. Even the oldest objects (in which 'heavy' elements constitute far less than one per cent of the total mass) turn out, however, to contain 23–24 per cent of helium. No star, galaxy or nebula has been found where helium is less abundant than this. It seems as though the Galaxy started not as hydrogen, but as a mix of hydrogen and helium. (The Sun has 27 per cent helium, the extra being just about what would have been made along with its carbon, oxygen and iron.)

Hoyle and his younger colleague Roger Tayler were the first to appreciate what this all-pervasive high helium abundance might signify. They suggested a different origin: an explosion far larger than a supernova. Hoyle had spent much of 1963 thinking about such phenomena. His interest had been aroused by the discovery of quasars, early that year (see Chapter 2). This led him to wonder whether huge superstars could exist, millions of times heavier than the Sun, and to calculate how much energy they might generate.

During the months when these thoughts were gestating, Hoyle was scheduled to give regular lectures in Cambridge, intended for advanced (graduate) students. The course had no predetermined syllabus; those who attended it in 1964 had the privilege of following, week by week, the emergence of ideas that are now canonical in the subject.

Hoyle and Tayler computed what would happen inside an *exploding* superstar weighing millions of times more than an ordinary star. If it started off hotter than 10 billion degrees, they found that roughly 25 per cent of its material converted into helium in the first 100 seconds; the exploding material would then become too cool and dilute to transmute the helium further up the periodic table. So, if everything had been 'processed' through a superstar, the helium mystery would be solved.

The quasars are now believed to involve huge black holes (see Chapter 5). But a consensus on this did not develop until the 1970s. When quasars were newly discovered, Hoyle and other theorists were encouraged in the belief that superstars might actually exist. But could there, plausibly, have been so many superstars that all the material in the universe had been processed through them? And if they are unstable, and explode so quickly, how could they have formed at all?

Such worries suggested a more radical picture. Perhaps helium was made in something even vaster than a superstar – a cosmic 'big bang' from which everything emerged. A year later, Penzias and Wilson discovered diffuse microwave radiation, which indeed seemed to be the afterglow of a hot 'beginning'. Hoyle then joined forces with Fowler, his old American colleague. They enlisted a younger collaborator, Robert Wagoner, to help them compute the entire network of nuclear reactions that could occur during the hot early phases. It was characteristic of his versatility and 'catholic' style that Hoyle, a die-hard advocate of the steady-state theory, also laid the foundations of a 'pillar' of the big-bang theory, which he then derided.

But can we really be serious about extrapolating back to when the universe was a billion times hotter than it is today? Atoms (or atomic nuclei) would have been 10^{27} (a billion cubed) times more closely packed than they are now. But our present universe is so diffuse – only about 0.1 of an atom per cubic metre – that even when multiplied 27 times by 10 the density is still less than that of air! Only if we extrapolate still further back, into the first *milli*second, do we need to worry about the uncertainties of ultra-high-density physics. And the nuclear reactions relevant to helium formation can be directly measured in the laboratory, and do not involve any large uncertain extrapolation from the experimental domain.

Helium is the only element that would be created prolifically in a 'big bang'. This is gratifying, because the theory of synthesis of elements in stars and supernovae couldn't explain why there was so much helium, and why the helium was so uniform in its abundance, even though it accounted very well for carbon, iron and so forth. Attributing helium to the Big Bang thus solved a long-standing problem, and emboldened cosmologists to take the first few seconds of cosmic history seriously.

The theory made a definite prediction that nothing should have less than about 23 per cent helium. Astronomers have made great progress in measuring the proportions of the elements in stars and nebulae. The helium abundance in the oldest objects is now pinned down to be 23 or 24 per cent. Another product of the Big Bang is *deuterium* (heavy hydrogen). An atom of deuterium contains not just a proton but a neutron as well, which adds extra mass but no extra charge. Deuterium has only a few hundred-thousandths of the abundance of hydrogen. But its origin poses a problem because it is destroyed rather than created in stars: as a nuclear fuel, it is easier to ignite than ordinary hydrogen, so newly formed stars quickly destroy their deuterium before settling down to their hydrogen-burning phases.

What is remarkable is that the abundances of helium and deuterium (and lithium as well) are all concordant with the predictions of 'Big-Bang nucleosynthesis'. These elements were made in the first few minutes when the universe was hotter than a billion degrees; the outcome (particularly the amount of deuterium) depends on exactly how dense the universe was then, which is directly related to how dense it is now. The measured abundances could have been entirely out of line with the predictions; or they might have been consistent, but only for a density that was (for instance) far lower than the universe actually has. As it turns out, the observed abundances all agree with what would be predicted, provided that there is, on average, 0.1 to 0.3 of an atom per cubic metre in the present universe. Gratifyingly, this is about the density that would result if all the material in all the galaxies were spread uniformly through space.[3]

SHOULD WE BELIEVE IN A HOT BIG BANG?

The hot-big-bang concept itself has more than fashion to commend it. There is real empirical support: it offers a consistent story about the history of matter and radiation.

The grounds for extrapolating back to the stage when our universe had been expanding for a second (when the helium began to form) deserve to be taken as seriously as, for instance, inferences from rocks and fossils about the early history of our Earth, which are equally indirect (and less quantitative). I would bet odds of ten to one that the hot-big-bang concept describes how our universe has evolved since it was around one second old. Some people are even more confident. At the International Astronomical Union's 1982 General Assembly in Greece, the Soviet cosmologist Zeldovich, who had, since the 1960s, contributed more theoretical ideas to the subject than anyone else, gave an exuberant lecture in an open-air theatre, in which he claimed that the big bang was 'as certain as that the Earth goes round the Sun'. He must, though, have known the dictum of his compatriot, the physicist Lev Landau, that cosmologists are 'often in error but never in doubt'!

Zeldovich died in 1987, only a few months after being allowed to make his first trip to the United States. (He had been, along with Sakharov and Kurchatov, a leader in the Soviet H-bomb programme. This involvement – though recognised by three awards as 'Hero of the Soviet Union' and no fewer than eight Orders of Lenin – led to his travel being even more restricted than that of his colleagues.)

The evidence that Zeldovich found so compelling is actually far stronger today. The putative relics of the big bang – the microwave background radiation and the so-called 'light elements' (helium, deuterium and lithium) – have been much more accurately observed. Moreover, one can think of several discoveries that *could* have refuted the model and which have *not* been made. For instance:

- Astronomers might have discovered an object whose helium abundance was zero, or at any rate well below 23 per cent, the absolute minimum that would emerge from the fireball (extra helium made in stars can readily boost helium *above* its pre-galactic abundance, but there is no way of converting all the helium back to hydrogen).
- The background radiation might have turned out to differ embarrassingly from the expected 'black body' or thermal form. In particular, the shortest wavelength (millimetre wave) intensity measured by COBE might have been *weaker* than the predicted extrapolation from what had already been reliably determined at centimetre wavelengths. Many processes could have added *extra* radiation at millimetre wavelengths – for instance, emission from dust, or from stars at very high redshifts. But it would be hard to interpret a millimetre-wave temperature that was *lower* than that at centimetre wavelengths.
- The 'fireball' would have contained *neutrinos* as well as photons. These particles interact very weakly with everything else, and would survive to the present time. There should be almost as many neutrinos as photons – the relative numbers are quite easy and unambiguous to calculate. There are now about 400 million photons per cubic metre; and there should be three kinds of neutrino, each with a density of 110 million per cubic metre. Neutrinos, therefore, outnumber the atoms by a huge factor – around a billion – just as the photons do. If each neutrino weighed even a millionth as much as an atom, they would, in toto, contribute too much mass to the present universe – more, even, than could be hidden in dark matter. Experimental physicists have been trying hard to measure the neutrino mass which is almost certainly very low. If they had got too high a positive answer, we would have had to abandon the idea of the big bang, but they did not.

The concept of a big bang has 'lived dangerously' for more than 25 years. Had various experiments and observations turned out differently, it would have been killed. The theory's survival gives us confidence in extrapolating right back to the first few seconds of cosmic history, and in assuming that the laws of microphysics were the same then as now. But we must remain alert to possible contradictions: our present satisfaction may reflect the paucity of the data rather than the excellence of the theory. Conceivably, this confidence is misplaced, and our satisfaction will prove as transitory as that of a Ptolemaic astronomer who has fitted a new epicycle.

When the steady-state theory was a serious contender, many cosmologists (not just its inventors) *hoped* observations would vindicate it. The idea was appealing because, if it were right, *all* key cosmogonic processes should be accessible to observation. In a steady-state universe, everything that ever happened – the origin of galaxies of all types, of all the chemical elements and so on – must be going on somewhere now. In a big-bang universe, on the other hand, the key features of our universe could be legacies of an early epoch shrouded from view, and frustratingly inaccessible to observation – so at least it seemed in those early days, when telescopes had only penetrated to modest redshifts.

But the 'steady statesmen' never expected that we could 'observe' very early epochs, and were therefore too pessimistic about the prospects of really understanding an evolving universe. The background radiation (and the elements such as helium) carry real quantitative information about the early phases. The key cosmogonic processes are proving just as accessible to observations, and as amenable to calculations and tests, as they could have been in a steady-state universe.

The microwave background probably ranks as the outstanding cosmological discovery of the last 50 years. (The only rival for the 'number-one slot' is the realisation of the extraordinary properties of black holes, and that such entities actually exist; this is introduced in the next chapter.) Ever since the 1960s, almost all cosmologists have been convinced that a big bang actually happened.

There is firm empirical support (and a firm link with 'known' physics) for inferences going back to when our universe was just a few seconds old – the implications from the microwave background, and from cosmic helium, have been the subject of this chapter. When later in the book, we venture right back into the first millisecond, we are on shakier ground, and shouldn't disguise this. Cosmologists shouldn't conflate things that are quite well established with those that are not yet in that state. When this distinction is blurred, there's a risk that readers will either accept speculations about the ultra-early universe overcredulously, or (if they are more sceptical) will fail to appreciate that some parts of cosmology, those pertaining to the later stages, are much more firmly grounded in observations, and in the physics that we can test in our laboratories.

Our present-day cosmic environment is a legacy of exotic physics in the first microsecond, when the mysteries of the cosmos and the microworld overlap. It isn't yet prudent to venture any bets on the details of such ultra-early eras. But some questions that were once entirely speculative are now coming into serious scientific focus. What determines the 'mix' of matter and radiation in the universe – the fact that there are a billion photons for every atom? Why does our universe have the overall uniformity that makes cosmology tractable, while nonetheless allowing the formation of galaxies, clusters and superclusters? And what imprinted the physical laws themselves?

You may be thinking: Isn't it absurdly presumptuous to claim that we can *ever* know anything about the beginnings of our entire universe? Not necessarily. It's complexity, and not sheer size, that makes a system hard to understand. The Sun is easier to understand than the Earth: it is hotter and denser, no minerals or chemicals could survive inside it, and everything is broken down into separate atoms. Likewise, in the even more extreme environment of the primordial fireball everything must surely have been reduced to its most basic components. The early universe could be easier to understand than the smallest living organism. It's biologists and the Darwinians who face the toughest challenge.

Boosted by this thought, I will press forward into more speculative territory. However, we must first discuss neutron stars, where matter is squeezed as dense as in the first millisecond of cosmic history; and black holes, where we confront, as in the big bang, conditions so extreme that they transcend our present knowledge. These objects, like our universe itself, are dominated by the force of gravity. The next chapter summarises what we've learned about this force, and the mysteries it still presents.

NOTES

1 Nor did they realise that there was already indirect evidence, dating back to 1941. Andrew McKellar, a Canadian astronomer, had been studying the spectra of stars. He found the telltale features of cyanogen (a molecule composed of one atom of carbon and one of nitrogen) in an interstellar cloud along the line of sight. He expected the cyanogen molecules to be completely cold, and in their lowest energy state (the so-called 'ground state'). But they were not: they seemed, from the properties of their spectra, to be bathed in radiation at (he estimated) 2.4 degrees. This work is recorded in Gerhard Herzberg's classic treatise entitled *Diatomic Molecules*; but Herzberg goes on to add that this temperature 'has only a very restricted significance'!

2 The Earth has lost its share of hydrogen and helium: these are volatile substances that could not be retained by the Earth's gravity. However, these elements are overwhelmingly dominant in the Sun and the giant outer planets. The other elements – carbon, oxygen, iron and so forth – have roughly the same proportions on Earth as in the Sun and most other stars.

3 These calculations constrain how many atoms in the universe can be hidden in some 'dark' form – this is relevant, as discussed in Chapter 6, to what the dark matter is, and whether there can be enough to bring the cosmic expansion to a halt.

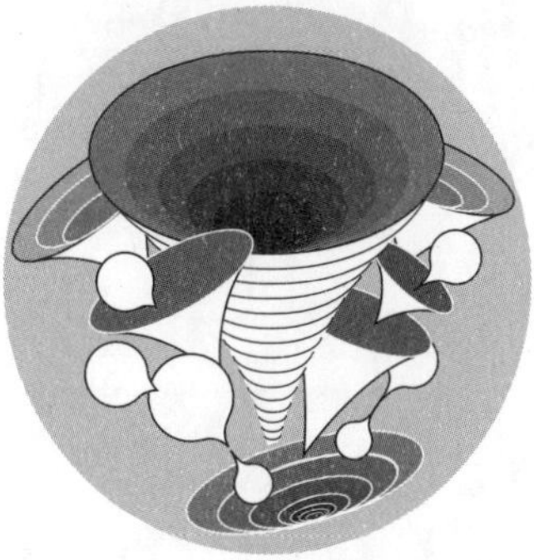

4

THE GRAVITATIONAL DEPTHS

Newton was not the first of the age of reason. He was the last of the magicians, the last of the Babylonians and Sumerians, the last great mind which looked out on the visible and intellectual world with the same eyes as those who began to build our intellectual inheritance rather less than 10 000 years ago.

John Maynard Keynes

FROM NEWTON TO EINSTEIN

Isaac Newton dedicated years of his life to alchemy and prophecy – as much intellectual effort, perhaps, as he ever devoted to gravity. He pursued his experiments in Trinity College, Cambridge. These sometimes related to physics – for instance, his famous experiments on light, using a prism. But chemistry preoccupied him: according to a contemporary account, 'for about 6 weeks at spring and 6 at the fall . . . the fire in the elaboratory scarcely went out'. In the words of John Maynard Keynes:

> He *did* read the riddle of the heavens. And he believed that by the same powers of his introspective imagination he would read the riddle of . . . past and future events divinely fore-ordained, the riddle of the elements and their constitution from an undifferentiated first matter, the riddle of health and of immortality. All would be revealed to him if only he could persevere to the end, uninterrupted . . . reading, copying, testing – all by himself. [Newton spent 25 years] in these mixed and extraordinary studies, with one foot in the Middle Ages, and one foot treading a path for modern science.

The 'modern' dimension of Newton's intense cerebration was his theory of gravity. His quest for a unified mathematical view of celestial motions was triumphantly successful, because the regularities of the Moon, planets and tides had already been chronicled for centuries. In contrast, an equally 'scientific' approach to chemistry would have been premature: realising how atoms and molecules underlie our everyday world would have been too great a leap for any 17th-century mind.

Newton envisaged gravity as a force transmitted instantly between stars and planets. Einstein has, however, taught us that no signal or influence can be transmitted faster than light: if, for instance, the Sun suddenly blew up, it would take eight minutes (the time light takes to reach the Earth) before we noticed anything amiss. This in itself implies that Newton's theory cannot be the 'last word'. But the planets move several thousand times slower than light, so Newton's laws are still adequate for almost all practical applications of gravity on Earth. They are even good enough to program a spacecraft's trajectory – to the moon, to Mars, or even on a 'grand tour' beyond Jupiter, passing through narrow gaps in Saturn's rings. These laws fail seriously only when gravity is far stronger than in the Solar System, and induces much faster motions.

Einstein's concept of gravity has greater scope. It encompasses situations where gravity is very strong, or the speeds very high; it tells us how gravity affects light itself; it allows us to describe an entire universe. Even more important, it affirms (rather than replaces) Newton's ideas by offering deeper insight into *why* all bodies fall at the same speed, whatever they are made of, and why an 'inverse square law' holds in the Solar System. Gravity seems less arbitrary and more natural; moreover, gravity is embedded in a conceptual scheme with extraordinary implications for the nature of space and time.

Einstein's general relativity predicts only very tiny deviations from Newtonian theory in the Solar System, but these have been confirmed by precise measurements. Experiments on the bending of light rays by the Sun's gravity have now been corroborated by tests using radio waves, but a thousand times more precisely. Experiments involving bouncing radar signals off planets, or tracking the orbits of space probes, are just as precise; and these too have vindicated Einstein's theory and refuted its rivals.

Remarkably, Einstein's theory wasn't stimulated by observations – it predated by 50 years the surge of astronomical discoveries that, since the 1960s, have revolutionised astrophysics and cosmology. In neutron stars and black holes, for instance, gravity is so strong that the modifications to Newton's laws are crucial, rather than being trifling corrections.

In 1905, the 26-year-old Einstein showed that, once we abandon the idea of 'absolute space', and accept that the laws of physics are identical in any steadily moving frame of reference, we must also abandon the idea that time can be measured in any 'absolute' sense – space and time are intimately connected. Einstein also, in the same year, proposed that light is quantised into 'packets' of energy (photons), and developed the theory of Brownian motion, the process whereby random impacts of molecules make specks of dust in a liquid jitter around. These contributions alone would rank him among the half-dozen great pioneers of 20th-century physics.

But it is his gravitational theory, 'general' relativity, promulgated ten years later, that puts Einstein in a class by himself. Even if he had contributed none of his 1905 papers it would not have been long before the same concepts were put forward by one or other of his contemporaries: the ideas were 'in the air'; wellknown inconsistencies in earlier theories and puzzling experimental results would, in any case, have focused interest on these problems. If Einstein hadn't existed, the insights in those early papers would surely have been reached within a few years, probably by three different people. What was astonishing was that they emerged, in a single year, from one unknown man working in the Berne Patent Office.

Einstein's vision of gravity in terms of curved space (so that 'space tells matter how to move; matter tells space how to curve') established him as the greatest physicist since Newton. This theory, almost ten years in gestation, was not a response to any particular observational enigma. It did indeed account for longstanding anomalies in Mercury's orbit, but Einstein was led to it by pure thought and deep intuition. He himself said when he announced his new work, 'Scarcely anyone who has fully understood the theory can escape from its magic.' Its internal logic seemed so compelling that he felt little need to defend it against criticism. Indeed, Einstein's new vision of gravity and inertia was widely (albeit tacitly) accepted even before its main consequences could be checked.

EINSTEIN'S VISION

Insight in science needs concentrated effort and preparation – that's true at the routine level, not just at the peaks scaled by Newton and Einstein. It also demands intuition and imagination. In this respect, it parallels artistic insight – equally an attempt to seek new patterns and new perspectives on the world. But those similarities shouldn't obscure one glaring difference between the two enterprises, which stems from the interlocking, cumulative and intensely social character of science. In the arts, individuality shines through even at the amateur level. Everyone's contributions, even if soon forgotten, are personal and distinctive. As Peter Medawar noted, when Wagner took ten years off in the middle of the *Ring* cycle to compose *Die Meistersinger* and *Tristan und Isolde*, he wasn't worried that someone would scoop him on *Götterdämmerung*.

Scientific advances, even when substantial and durable, generally merge almost anonymously into the whole edifice of 'public knowledge'. Scientists' personal imprints may fade; but they have the compensation that, if it survives critical scrutiny, their work endures. And they have a second compensation. Critics of literature or painting are seldom themselves creative artists; but every researcher is, albeit in a small way, both a creator and critic of the collective scientific enterprise.

Discoveries generally emerge when the time is ripe. Individuals seldom make more than a few years' difference to when a particular advance occurs. There are exceptions – Charles Townes, co-inventor of the maser, has for instance claimed that masers could have been made much earlier; and, to take a quite different example, the concepts of sociobiology could have been fully developed in the 1920s. Einstein is unique among 20th-century scientists in the degree to which his work retains its individual identity: without him, we might have had to wait decades for equivalent insights on gravity. This gives special interest to his personality and style.

Documentation of Einstein's life was, until recently, remarkably meagre for such an illustrious figure. He published an autobiography, but this eschewed almost entirely what he called the 'merely personal'. After his death in 1955, his literary executors were reluctant to release material that might detract from a romanticised image. Publication of his collected writings was beset by delays. The highlight of the first instalment, which didn't appear until 1987, was Einstein's correspondence with Mileva Maric, who became his first wife. She was a fellow student, of Hungarian origin, at the Federal Polytechnic in Zurich, one of the few first-rate scientific centres then open to women.

The young Einstein had no immediate success in getting a regular academic job. He had antagonised his former professor; antisemitic prejudice was a further handicap. After two years of casual tutoring, his appointment as a 'technical expert, third class' in the Berne Patent Office gave him some security. Einstein's letters to Mileva are interspersed with comments about physics; sometimes he refers to 'our work'. Most scholars discount the claim that she should rank as co-discoverer of relativity (which, if true, would not have been the first case of a women scientist being denied due recognition; nor, sadly, the last). The prime 'sounding board' for the ideas in Einstein's early papers was Michele Besso, who remained a lifelong friend; they set up an informal discussion group, facetiously called the Olympian Academy. (Einstein recalled it as 'far less childish than those respectable ones which I later got to know'.) But his isolation from academic physics was never complete. Nor was it prolonged: he was himself a professor when, in his early thirties, his theory of general relativity was gestating.

Even the greatest scientists seldom become celebrities. Throughout the years of his peak creativity, Einstein was little known to the non-scientific public. But his status was transformed in 1919, when measurements made during an eclipse apparently confirmed that light rays were slightly deflected, as he had predicted, by the Sun's gravity. The 'prime mover' behind these observations was Arthur Eddington, a Cambridge astrophysicist best known for his researches on the nature of stars, who was among the first to understand and appreciate the new theory.

Media interest in science is always capricious, but this arcane astronomical result triggered extraordinary press hype: NEWTONIAN IDEAS OVERTHROWN was *The Times*'s headline, followed two days later by LIGHTS ALL ASKEW IN THE HEAVENS: MEN OF SCIENCE MORE OR LESS AGOG . . . EINSTEIN THEORY TRIUMPHS in the *New York Times*.

Einstein's life could thereafter never be the same; he was constantly in the public gaze. In Germany, where he remained until the early 1930s, he confronted attacks on 'Jewish science'. On the international front, he espoused Zionism, the League of Nations, and (in his last years) nuclear disarmament. He seemed to relish fame, happy to be pictured with Charlie Chaplin and other celebrities of the time. It was fortunate for science – not just for journalists seeking ready quotes – that its pre-eminent practitioner purveyed such an engaging and idealistic image.

In 1936, Einstein was irrevocably exiled from Europe; the condition of his son Edouard, a schizophrenic, was worsening. At this bleak time he wrote that 'as long as I am able to work I must not and will not complain, because work is the only thing which gives substance to life'. Stoicism is admirable only if the misfortune is as deeply felt as it should be, and the depth of Einstein's emotional engagement may indeed sometimes have been in question. But, without such detachment, Einstein would never have sustained the concentration his work required. To achieve his insights, he needed, like Newton, to 'think on them continually'.

Newton's later years were spent in public office as 'Master of the Mint', as a dignitary rather than a researcher (though his chemical experiments may have proved not entirely irrelevant). In contrast, Einstein's intellectual motivation never flagged – he worked and calculated until his last days. Some of Einstein's ideas invoked a fifth dimension (in addition to time, and the three dimensions of ordinary space) to represent electric and magnetic forces. He was, in effect, seeking what would now be called a unified theory. The forces that hold atomic nuclei together hadn't even been discovered then, so in retrospect this quest was plainly premature (as indeed it perhaps still is), and it sidelined him from 'mainstream' physics in the last 30 years of his life.

The familiar image, the icon of posters and T-shirts, depicts the old Einstein, a benign and unkempt sage – a contrast to the boisterous and wilful young man who revolutionised physics.

Richard Westfall, in the preface to his classic biography of Isaac Newton, wrote that, the more he studied him, the more 'wholly other' he seemed. Einstein ranks perhaps second only to Newton in the pantheon of scientific intellects, but his personality seems far less alien. His recently published letters and papers redress the sanitised blandness of earlier biographies, and illuminate the private life that was the backdrop to his extraordinary achievements.

Because general relativity was put forward so far in advance of any real application, it remained, for 40 years after its discovery, an austere intellectual monument, a somewhat sterile topic isolated from the mainstream of physics and astronomy. This is in glaring contrast to its more recent status as one of the liveliest frontiers of fundamental research. Its reputation for being an exceptionally hard subject was always exaggerated. (Arthur Eddington was asked if he was really one of the only three people in the world who understood general relativity. He is rumoured to have hesitated in his response, not through modesty, but because he couldn't think who the third might be.) But it is now taught almost routinely in universities, along with quantum mechanics, electromagnetism and the rest of the 'menu' offered to physics students.

But where in our universe is gravity strong enough to reveal the distinctive consequences of Einstein's theory? It is worth addressing this question before delving further into the theory itself.

STRONG GRAVITY IN NATURE: NEUTRON STARS

Inside the Sun, gravity's inward pull is balanced by the pressure in its hot core. If the central pressure were removed, the Sun would go into free-fall, halving its size in less than an hour. If, on the other hand, gravity were magically switched off, the hot interior would just as suddenly explode and disperse. The Sun is (like most other stars) almost in equilibrium. Pressure balances gravity, and nuclear fusion in the core generates just enough energy to replenish the heat lost from its surface.

Heavier stars expend their energy more quickly, and end their lives as supernovae – the explosions whose debris, scattered through interstellar space, is crucial in the stellar recycling process that synthesised all the elements of the periodic table.

Before anyone recognised the role of supernovae in the 'ecology' of galaxies (described in Chapter 1), there were speculations about how stars explode, and what remnants they may leave behind. The first correct conjecture about supernovae dates back more than 60 years. In a short paper published in 1934, Walter Baade, an astronomer at the Mount Wilson observatory, and his colleague Fritz Zwicky wrote, 'With all reserve we advance the view that a supernova represents the transition of an ordinary star into a neutron star, consisting mainly of neutrons.' They speculated that a supernova explosion was driven by the gravitational energy impulsively released when the star's core collapsed, and that a tiny cinder should remain. The heavy central nuclei of atoms were already known to be tiny compared with the atoms themselves, whose overall dimensions (and spacing in ordinary solids) are set by the diffuse 'cloud' of electrons surrounding the nucleus. In a 'neutron star' Baade and Zwicky postulated that the nuclei themselves were closely packed. The entire content of a star could then be squeezed within a radius of 10 kilometres – millions of times denser even than white dwarfs. The volume of a sugarlump could contain a hundred million tonnes of neutron star material.

Why neutrons? The nuclei of ordinary atoms are made up of protons and neutrons. For example, helium has a nucleus of two protons (each with a positive charge) plus two neutrons (with no electric charge); iron has 26 protons and 30 neutrons. In the laboratory, an isolated neutron is unstable, and decays into a proton and an electron. On the other hand, at extreme densities the process goes the other way: protons turn into neutrons.

Zwicky was the first astronomer to search systematically for supernovae in other galaxies, and to classify them into different types. He continued to speculate about neutron stars, but, despite his remarkable insights, he did not understand enough physics to work out the details. One person who certainly had the expertise was Robert Oppenheimer. Before his celebrated leadership of the 'Manhattan project' to develop the atomic bomb, Oppenheimer led a lively research group at the University of California; with his student George Volkoff, he used the best contemporary knowledge of atomic nuclei to calculate what a neutron star would be like.

Despite this theoretical interest in the late 1930s, Baade and Zwicky's conjecture remained just that right until 1968, when an ordinary-looking little star in the middle of the Crab Nebula was actually found to be flashing on and off thirty times a second. Nothing but a neutron star could be compact enough to spin so fast without flying apart.

PULSARS

The 'remnant' in the Crab Nebula was not the first neutron star to be discovered. Priority went to Anthony Hewish and Jocelyn Bell, Cambridge radio astronomers, whose discovery of 'pulsars' constituted one of the most remarkable pieces of serendipity in modern science.

Hewish built a special instrument with an important special feature: it was sensitive enough to record *rapid changes* in the intensity of the radiation from distant sources.[1] And he found what he was looking for: just as stars twinkle because their light passes through turbulent air, so some radio sources 'scintillate', because the radio waves pass through an irregular medium on the way towards us. But his research student, Jocelyn Bell, found variations of a quite distinctive kind – sporadic series of regular pulses, each pulse lasting a fraction of a second, coming from specific points in the sky. A frantic few months of effort ensued. The Cambridge radio astronomers had to check whether the signals had a terrestrial origin (maybe some secret space project?). Three more of these mysterious sources were soon found, each ticking at a well-defined rate. Could they perhaps be signals from intelligent extraterrestrials? This idea was never taken very seriously, but the sources were jocularly referred to as LGM 1, 2, 3 and 4 (for 'little green men').

When this discovery was announced in the journal *Nature*, even the other astronomers in Cambridge were astonished. Hewish and his colleagues had not shared their excitement with anyone outside a tight-knit group. This concealment annoyed some of us at the time, but in retrospect I think Hewish was no more than prudent.

Only a few months elapsed between Jocelyn Bell's first intimations and the actual publication, so nobody's chance of follow-up work was seriously delayed. And, for most of those months, Hewish and Bell weren't completely confident that the signals were 'real'. If the sporadic radio pulses had turned out to have a mundane interpretation, or to arise from some fault in their equipment, a premature announcement would not only have been embarrassing, but might have wasted the efforts of many other astronomers who would undoubtedly have followed up any rumour of this kind.[2]

What could these objects be? An ordinary star like the Sun would fly apart if it pulsed or rotated much faster than once per hour. Bodies that turned on and off in a fraction of a second plainly had to be much more compact. Were they white dwarfs, or maybe neutron stars? Were they pulsing or spinning? All these options (and many others) had their advocates. The Cambridge group originally favoured pulsating white dwarf stars. (A naïve enquirer at a press conference was perplexed about how, at such a great distance, white dwarfs could be distinguished from little green men!)

The case for rotating neutron stars was first clearly argued by Thomas Gold (mentioned in Chapter 2 as co-inventor of the steady-state cosmology; he had by this time moved to Cornell University in the US). There were good reasons for expected neutron stars to form when the cores of heavy stars collapsed, triggering supernova explosions. They would be so small, and have such strong gravity, that they could spin as fast as a thousand revs per second without flying apart. The spin rate would provide a natural stable clock; a 'lighthouse beam' anchored to the star, would send an intense pulse towards us once per revolution.

Only a year later, the debate was settled in Gold's favour. A very fast pulsar was found at the centre of the Crab Nebula, transmitting 30 pulses per second: a white dwarf could neither rotate nor pulsate as fast as that, but such rapid spin was no problem for a neutron star. Moreover, careful timing showed that the pulse rate was gradually slowing down: this was natural if energy stored in the star's spin was being gradually converted into radiation, and into a wind of particles which keep the Crab Nebula shining in blue light.

Baade and Zwicky's early speculations (see page 74) were fully vindicated. Zwicky especially enjoyed quoting a succinct summary of his ideas in a cartoon strip, 'Be Scientific with Ol' Doc Dabble', in the *Los Angeles Times*. In January 1934 this featured the caption,

'"Cosmic rays are caused by exploding stars which burn with a fire equal to 100 million suns, and then shrivel from ½ million miles diameter to little spheres 14 miles thick," says Prof. Fritz Zwicky, Swiss Physicist'. More than 30 years later, Zwicky could justly say, 'I told you so.'

NEUTRON STARS

Even scientists who feel no urge to explore the cosmos for its own sake are interested in the parts of our environment where conditions are specially extreme. Supernova explosions, and the neutron stars they leave as remnants, are fine instances of this.

Neutron stars offer a fascinating 'cosmic laboratory' for studying conditions far more extreme than could ever be simulated on Earth. A slice through a neutron star would resemble a slice through the Earth, with a crust, a liquid interior and maybe a solid core. The outer crust would be mainly iron; deeper down, the pressure is so high that individual nuclei merge into a 'neutron fluid' with the same rather unusual properties as helium at temperatures of a few millidegrees above absolute zero – a 'superfluid', which can flow without any resistance at all.

At first sight, it might seem hopeless to check our ideas about the inside of neutron stars when we are so remote from them. But, amazingly, we can. Pulsars can be timed to an accuracy better than a microsecond. On average, they are slowing down, but there are occasional 'glitches', when they suddenly speed up by a tiny amount. There is one straightforward reason for expecting 'glitches'. Just as the Earth isn't an exact sphere, but bulges outward at its equator, so does a spinning neutron star. As its spin slows down, the equatorial bulge diminishes. If the star were entirely a fluid, the shape would readjust continuously. But, because the crust is rigid, stress builds up until there is a sudden crack – a starquake. The size and frequency of these quakes tell us how thick and rigid the crust is. A movement of even a few microns (thousandths of a millimetre) perturbs the star's spin rate by enough to show up clearly in the measurements. It is astonishing that such microscopic effects are noticeable in a star thousands of light-years from us.

These glitches have been studied in detail, and it is now believed that the most common ones result from a different effect: namely slippage between the solid crust and the fluid core. The drag that slows down a pulsar's rotation is applied to the crust, and doesn't directly slow down the liquid core. The core then rubs against the slower-spinning crust, and is slowed down by friction. This force doesn't act smoothly, however: instead, it operates in jerks rather like a worn-out clutch, and, whenever the friction suddenly increases, the crust speeds up.

The precise study of how their spin-rate changes is just one of the ways we can learn about neutron star interiors – astrogeology, unlike astrology, is a feasible and serious subject.

The aspect of neutron stars that remains most perplexing even today is, ironically, the intense radio emission that enabled Hewish and Bell to discover them in the first place. Almost certainly a magnetic field is implicated – like the Earth's, but many billion times stronger. But, again as with the Earth (where 'magnetic north' is not the same as true north), the magnetic axis is misaligned with the spin axis. The radio emission, linked to the magnetic-field orientation, therefore sweeps around the sky, beaming towards us, and giving a 'pulse' once per revolution.

Nearly a thousand pulsars are now known. Each is a flashing beacon, marking the explosive death of a massive star that happened (in most cases) millions of years ago. Most are spinning much more slowly than the Crab pulsar, consistent with having slowed down as they have aged.[3] The ejected debris from the supernova explosions – the 'nuclear waste' from their precursor stars – would long ago have dispersed into the interstellar medium.

Neutron stars are also extreme in the strength of their *gravity*. Its force is 10^{12} times that on Earth. Their surfaces would be smooth, with no mountain more than a millimetre high. But so strong is gravity that more energy would be expended in climbing a millimetre mountain on a neutron star than in escaping completely from the Earth. Here on Earth, a pen dropped off a table just makes a noise; dropped from the same height onto a neutron star, it releases as much energy as a tonne of high explosive. To escape completely from a neutron star would required a rocket moving at about half the speed of light (150 000 kilometres per second, compared with 11 kilometres per second to escape from the Earth).[4]

In our Solar System, light rays are only very slightly bent by gravity – exact measurements are needed to find this effect at all. But near a neutron star the 'light bending' would be 10–20 degrees – enough seriously to distort the view. No structure would escape squashing on the surface of a neutron star, so the only possibility of ever observing such effects would be from an orbiting spaceprobe. Even from that vantage point, however, you would be stretched uncomfortably: gravity gets weaker with height, and the difference between the gravitational pull on the top and bottom of the spaceprobe (the 'tidal' force) would be large.

Before leaving pulsars, I'd venture two speculations – two historical 'might-have-beens'. The pulsar in the Crab Nebula can be seen through any large telescope. It emits pulses of visible light as well as radio waves. But the pulse repetition rate, 30 per second, is so high that the eye responds to it as a steady source. Had it been spinning more slowly – say 10 times a second – the remarkable properties of the little star in the Crab Nebula could have been discovered 70 years ago. How would the course of 20th-century physics have been changed if superdense matter had been detected in the 1920s, before neutrons were discovered on Earth? One cannot guess, except that astronomy's importance for fundamental physics would surely have been recognised far sooner.

Another 'near miss' came in 1964, when Hewish and a research student from Nigeria, Sam Okoye, unknowingly detected the pulsar in the Crab Nebula. They did not actually record pulses, but they proved that the radio emission from the middle of the Crab Nebula came from a new kind of source smaller than any other known at that time. By following this up, they might have found the pulses. Had history gone that way, the Crab pulsar would have been the first neutron star to have been discovered. Hewish would still have been co-discoverer of pulsars, but four years earlier, and with Sam Okoye rather than Jocelyn Bell.

X-RAYS FROM NEUTRON STARS

Neutron stars were found by a lucky accident. No one expected them to be pulsars, with strong and distinctive radio emission, so there had been no systematic radio searches for them. If theorists in the early 1960s had been asked the best way to detect a neutron star, most would have looked blank; those who did not would have suggested a search for *X-ray* emission. The reasoning here is straightforward. If neutron stars radiate as much energy as ordinary stars, but from a much smaller surface, they must be far hotter – hot enough to emit, not in the blue, nor even in the ultraviolet, but in the kind of radiation known as X-rays. To the physicist, X-rays (like radio waves, light and ultraviolet radiation) are a form of electromagnetic waves, but with shorter wavelengths and faster, more energetic, oscillations. So it was *X-ray astronomers*, not radio astronomers, who had been *expecting* to discover neutron stars.

X-rays from cosmic objects get absorbed by the Earth's atmosphere, and so can be observed only from space. X-ray astronomy, like radio astronomy, received its impetus from wartime technologies and expertise. But in this case it was scientists in the US who took the lead, especially Herbert Friedman and his colleagues at the US Naval Research Laboratory. The first X-ray detectors, mounted on rockets, each yielded only a few minutes of useful data before they crashed back to the ground.

Friedman's group carried out a famous experiment in 1964. They fired a rocket just at the time when the Crab Nebula was about to be occulted by the Moon. If the X-rays came from the whole nebula, they would fade gradually as the Moon moved across it; if, on the other hand, they came from a central point source (a neutron star?) they would be extinguished suddenly. Friedman observed a *gradual* fading; this told him that the X-rays came mainly from the same glowing material that shines diffusely in blue light. We now know that some X-rays do come from the pulsar, but Friedman was unlucky that these amounted to only 10 per cent of the total emission – his equipment wasn't sensitive enough to detect them.

X-ray astronomy derived its impetus from superpower rivalry in space technology and nuclear weapons. But it spurted forward in 1970 when NASA launched the first X-ray satellite, which could gather data for years rather than just for a few minutes. This small satellite was built and operated by a research group led by Riccardo Giacconi, an Italian physicist who had settled in the US. These pioneers of a new field of astronomy actually worked in a company called American Science and Engineering, whose main 'customer' was the US Defense Department: their first instruments were built to detect X-rays from thermonuclear tests. Giacconi's group later moved to Harvard University and the Smithsonian Observatory, where they developed larger X-ray telescopes; his old company reverted to more mundane products such as baggage-scanners for security checks at airports.

Giacconi's satellite was launched from a site in Kenya, and given the name *Uhuru* (the Swahili word for 'freedom'). *Uhuru*'s intended purpose was to discover X-ray sources beyond our own Galaxy. In that respect (and in that only) the project fell short of expectations: fewer clusters of galaxies and quasars were found than had been predicted. Its main achievement was to detect entirely unexpected X-ray emission from within our Galaxy. Repetitive pulses, with periods of a few seconds, came from objects in close, almost grazing, orbits around relatively ordinary companion stars. These sources are, like the radio pulsars, spinning neutron stars. However, their radiation is produced very differently. Gas captured from the companion star is pulled towards the neutron star by its strong gravity, impacting on the surface at more than half the speed of light. The energy released in the impact is radiated as X-rays. The inflow is channelled by the magnetic fields, so that gas lands preferentially near the magnetic poles. The observed X-rays come from 'hot spots' around the magnetic poles, and are therefore modulated with the star's rotation period.

The actual chronology whereby neutron stars were discovered seems, in retrospect, very arbitrary: the pattern could easily have fallen into place in a different order. Had Friedman's 1964 experiment been slightly more sensitive, he would have found that 10 per cent of the Crab Nebula's X-rays do indeed come from a central point source, and the first neutron star to be discovered would have been that in the Crab. (This could have happened even before Hewish and Okoye's 'near miss'.) On the other hand, had Jocelyn Bell been less perceptive, neutron stars might have escaped notice until the launch of *Uhuru* – the pulsing objects in binary systems would then have been quickly interpreted as neutron stars.

CHECKING EINSTEIN'S THEORY

Einstein's general relativity is now bolstered by firm evidence. Until the 1960s, no objects were known whose gravity was 'strong'. And the experiments seeking small departures from Newton's theory in the Solar System were so imprecise that several alternative theories were still in the running. But now, the orbits of planets have been measured more precisely thanks to radar; artificial spaceprobes can be tracked even more accurately; and radio astronomers can improve on optical measurements of light-bending by the Sun's gravity. These techniques have confirmed general relativity to better than 1 part in 1000, and eliminated most of its rivals.

An even better 'test bed' was discovered by Richard Hulse and Joseph Taylor in 1974: the remarkable 'binary pulsar'. In this system, a pulsar is orbiting around another neutron star once every eight hours. Because this is a much smaller and tighter orbit than those of planets in our Solar System, the deviations from Newtonian gravity are more substantial. According to Einstein's theory, the orbit of Mercury should precess by a tiny (but just measurable) amount. The analogous precession in the orbit of this binary pulsar is 10 000 times faster. Taylor has been making increasingly precise measurements of this system for more than 20 years. He has not only tracked the precession, but discovered that the orbit is gradually shrinking. This confirms the phenomenon of 'gravitational radiation', another prediction of Einstein's theory. Any moving system whose gravity changes generates tiny vibrations in space itself, which carry away energy. This effect, far too small to be detected in our Solar System, shows up clearly in the 'binary pulsar'.

Einstein's theory can be tested in further ways: for instance, scientists at Stanford University have built an extremely precise gyroscope which they hope to place in orbit around the Earth to detect a tiny but distinctive precession effect that relativity predicts. This experiment was first proposed in the 1960s, and the case for doing it was then very strong. Paradoxically, the case has got weaker as the evidence for general relativity has strengthened. If the experiment gets a result that agrees with Einstein, most people will be unsurprised. But, if it gets a discrepant result, few people would abandon their belief in Einstein's theory; most would, instead, suspend judgement until this novel and technically challenging experiment had been independently repeated. So the most exciting outcome would be a request for funds to repeat the whole thing!

The accurate verifications of the small deviations from Newtonian theory, the so-called post-Newtonian effects, enhance our confidence that Einstein's theory is valid. One still awaits, however, a direct astronomical diagnostic of the exact behaviour of gravity when its effects are very strong.

PLANETS AROUND A PULSAR

Pulsars are tiny 'dead' stars spinning very fast, which emit a beam of intense radio waves that points towards us once each revolution. The resultant regular pulses can be timed very accurately. This great precision allowed Taylor to test Einstein's theory. It has also enabled another radio astronomer, Alex Wolzczyan, to discover planets orbiting around a pulsar.

As mentioned in Chapter 1, planets around *solar-type* stars completely eluded astronomers until 1995. But a planetary system around a pulsar was found three years earlier. Wolzczyan had been using the giant radio dish in Arecibo, Puerto Rico, to monitor one particular pulsar for several years, and found irregularities in the arrival times of its pulses, which meant that the pulsar's position was wobbling slightly. His remarkable breakthrough was to recognise that these irregularities were the combined effects of two (or possibly three) planets orbiting the pulsar. These planets, smaller than our Earth, are exposed to intense radio waves and streams of fast particles from the pulsar, rather than to ordinary starlight, and would be unpropitious for life. This unusual planetary system nevertheless offers new clues to how planets form in general. Wolzczyan succeeded because the techniques radio astronomers use to detect tiny changes in the motion of a pulsar – measuring the arrival time of pulses with accuracies of a microsecond – are very much more sensitive than any that optical astronomers can yet deploy on normal stars.[5] The latter have so far only detected planets like Jupiter, hundreds of times heavier than our Earth.

THE TECHNICAL IMPETUS

The nature of gravity is one of the 'fundamental' issues of science – the deepest insights stem from one man, Einstein, who ranks in the public mind as the archetype theorist. But, even in this field of enquiry, observers and experimenters have made much of the running – progress depends on their technical originality and persistence. Martin Ryle's achievements in radio astronomy, for instance, stemmed from his inventiveness in electrical engineering. And the other key figures in this chapter – Herbert Friedman, with his background in rocketry; Riccardo Giacconi, who developed X-ray detectors, and Joseph Taylor, whose timing of pulsars is accurate to 15 decimal places – have contributed through their technical brilliance. Theorists may seem to focus more directly than their co-workers on the goal of the scientific enterprise – to interpret and understand what is discovered. But the insights that lead to technical innovations are often the real 'drivers' of progress. (The intellectual leap involved in inventing a zip fastener far surpasses what many theoretical physicists ever achieve!)

As we've seen, it was a matter of luck that the credit for discovering neutron stars went to serendipitous radio astronomers rather than to X-ray astronomers who were expecting them, and actively seeking them, in the 1960s, and who did in fact discover them by a quite different route three years later. But X-ray astronomers have undoubtedly taken the lead in confirming the most remarkable prediction of Einstein's theory – black holes.

NOTES

1 Hewish was primarily interested in the peculiar distant galaxies that emitted so powerfully in the radio-frequency band. These were the objects whose statistical properties enabled Martin Ryle to strike the first blow against the steady-state theory (see Chapter 2). Early radio telescopes, unfortunately, gave a blurred view: they did not reveal, for instance, whether the radio emission came from deep inside a galaxy or from a fuzzier volume surrounding it. Hewish invented a special technique for diagnosing the size of a radio source. His method exploited the same physical principle that causes stars to twinkle whereas planets do not: starlight is irregularly refracted in the upper atmosphere, but the irregularities are so small that a planet, whose image is an extended disc, covers so many of them that the effect averages out. Hewish discovered that the diffuse gas flowing out from the Sun into interplanetary space, the 'Solar Wind', affects *radio* waves rather as the upper atmosphere affects visible light. If radio sources were billions of light-years away, as Ryle believed, then they would 'twinkle' or scintillate if they were smaller than a galaxy; if they were even larger and fuzzier than galaxies, they would not.

2 Jocelyn Bell received less than her fair share of credit for the discovery of pulsars. This happened, I think, because of the social pressures which (then even more than now) impeded women's careers, and lowered their scientific aspirations. After getting her PhD, Jocelyn Bell left active research for several years – giving priority to her husband's career seemed at that time the 'natural' thing to do. Had she instead continued, and acquired

'visibility' by joining the small cohort of radio astronomers who, over the next few years, consolidated our knowledge of pulsars and discovered many more – as, almost certainly, a *man* with her extraordinary initial record would have done – it is hard to believe her achievements would have been slighted to the same extent.

3 Some pulsars, however, spend part of their life cycle in orbit around a companion star. These have more complicated histories, and sometimes have been 'spun up' to even faster rotation rates – up to 600 revolutions per second. These 'millisecond pulsars' have weaker magnetic fields than the others, for reasons that are still unclear. However, they have much smaller glitches in their spin-rate, and are therefore even better clocks. One inference that we can draw from the steadiness of these natural clocks is described in Chapter 12).

4 The energy that would have to be supplied in order to disperse the star (the so-called 'gravitational binding energy'), is typically 20 per cent of the rest mass energy (mc^2). Indeed, it is from gravitation that the magnetic and rotational energy of pulsars is derived. Ordinary stars are slowly spinning (the Sun's spin period, for instance, is about a month). When a star (or its core) collapses to the size of a neutron star it spins much faster and more energetically. This spin energy leaves a 'dead' star with more energy than nuclear reactions could generate over its entire previous lifetime. It is this energy, stored as though in a huge flywheel, and ultimately derived from gravity, that keeps pulsars shining.

5 A few months before Wolzczyan's announcement, a radio astronomer at Jodrell Bank in England, Andrew Lyne, had claimed (on the basis of similar arguments) that there was a planet around another pulsar. He turned out to be mistaken: the irregularities in the pulse arrival time that Lyne attributed to a planet actually arose from a subtle mistake he made when correcting for the Earth's motion round the Sun. However, had Wolzczyan not known about Lyne's claim, he might not have scrutinised his own data so carefully, and to such remarkable effect – an example, perhaps, of how even wrong work can sometimes be a constructive stimulus.

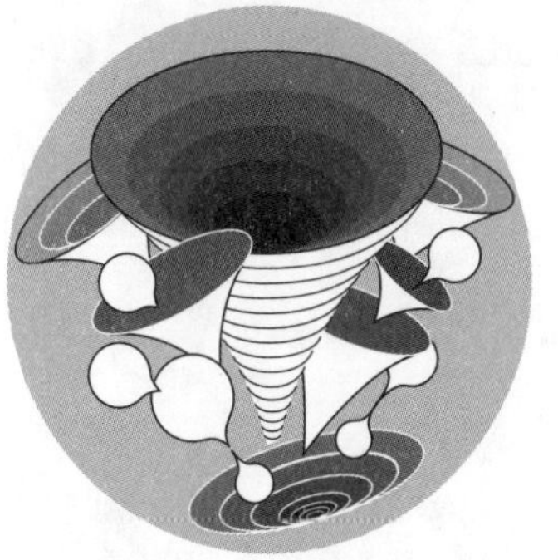

5

BLACK HOLES: GATEWAYS TO NEW PHYSICS

I became possessed with the keenest curiosity about the whirl itself. I positively felt a wish to explore its depths, even at the sacrifice I was going to make; and my principal grief was that I should never be able to tell my old companions on shore about the mysteries I should see.

A Descent into the Maelstrom, E.A. Poe

SOME PREMONITIONS

Gravity is strong in neutron stars, but there are other bodies where it is even more overwhelming. These are 'black holes' – bodies that have collapsed so far that no light, nor any other signal, can escape from them. They are expected in many theories of gravity;

not just Einstein's. Indeed they were in essence conjectured over 200 years ago. In 1783 John Michell presented a paper to the Royal Society of London about how gravity might affect light. He was an underappreciated polymath, a clergyman who also studied experimental physics and binary stars (see Chapter 2).

Michell noted that a projectile thrown from the surface of the Sun would need a speed about one-five-hundredth that of light if it were to escape. He realised also that the 'escape velocity' (as we would now call it) could be much larger for a body heavier than the Sun – in fact a simple calculation based on Newtonian gravity shows that the escape velocity, for bodies of a given density, is proportional to their radius. Therefore, he said,

> If the semi-diameter of a sphere of the same density as the Sun were to exceed that of the Sun in the proportion of five hundred to one, and supposing light to be attracted by the same force in proportion to its *vis inertiae* with other bodies, all light emitted from such a body would be made to return towards it, by its own proper gravity.

So Michell was suggesting that the most massive might be undetectable by their direct radiation, but may still manifest gravitational effects on material near them.

Pierre Laplace advanced the same idea in 1794, in his book *Le Système du Monde*. He actually deleted this particular passage from later editions: it is not clear why – probably he had, in the meantime, lost confidence in an argument that envisaged light as 'ballistic' particles.

Newton's theory is inadequate when the motions induced by gravity approach the speed of light. But general relativity copes quite consistently with situations when gravity is overwhelmingly strong, as it is in black holes – indeed it is there that Einstein's equations have their most remarkable consequences.

BLACK HOLES ACCORDING TO EINSTEIN

Less than a year after Einstein announced his theory, the German astronomer Karl Schwarzschild applied it to calculate how gravity behaved around a spherical mass. Einstein had already worked out how light would be deflected near the Sun; but this required only an approximate calculation, because gravity in the Solar System is weak enough that the bending is small. He knew his equations were hard to solve exactly, and was therefore enthusiastic that this solution had been found. Schwarzschild died shortly afterwards, from a disease contracted while serving as an artillery lieutenant in the German army.

Schwarzschild did his calculations more than 50 years before neutron stars were discovered. These were the first objects where gravity was strong enough to reveal the distinctive features of Einstein's theory. The passage of time is distorted: a clock on the star's surface would seem to a distant observer to be ticking more slowly. Associated with this is the 'gravitational redshift': radiation from the surface reaches distant observers with a lower frequency (and longer wavelength). These effects may be as large as 30 per cent on the surface of a neutron star.

What would happen if a neutron star were to contract further, or more mass were squeezed into the same radius? Gravity would then become so strong that not even light could escape its pull. The minimum radius from which light could escape happens – though this is little more than a coincidence – to be exactly equal to what Michell calculated. At this 'Schwarzschild radius' the gravitational redshift is infinite. It is, in effect, a 'horizon' shrouding the interior from view, a semipermeable membrane that can be crossed only in the inward direction.

Light rays near a black hole are even more strongly curved than around a neutron star. An experimenter situated just outside the 'horizon' would need to aim a beam of light almost exactly radially in order that it should not be bent back and eventually swallowed by the black hole. No one who ventured within could send *any* light signals to the external world – it is as though space itself is being sucked inward faster than light moves out through it. A freely falling astronaut could pass inward without experiencing anything especially unusual on crossing the horizon. But there would then be no chance of escape. An external observer would never witness the astronaut's final fate: any clock would appear to run slower and slower as it fell inward, so the astronaut would appear impaled at the horizon, 'frozen' in time.

The Russian theorists Zeldovich and Novikov, who studied how time was distorted near collapsed objects, coined the term 'frozen stars'. The term 'black hole' was not coined until 1968, when John Wheeler described how an infalling object 'becomes dimmer millisecond by millisecond ... light and particles incident from outside ... go down the black hole only to add to its mass and increase its gravitational attraction'.

Most things in our universe are spinning, and are consequently not spherical. The Schwarzschild solution is therefore too 'special' to apply to real objects. In 1963 the New Zealander Roy Kerr discovered a more general solution of Einstein's equations, which represented a collapsed *rotating* object. As in the Schwarzschild case, a horizon shrouds the exotic interior from view. But the space behaves like a vortex: objects are forced to swirl around, as well as (if they get too close) falling inexorably inward.

The hole's interior cannot be observed from a safe distance. One's Faustian urge must be sufficiently strong to venture inside! What would then happen? An infalling astronaut would experience stronger and stronger tidal forces (the difference between the gravitational acceleration of hands and feet); these forces diverge towards infinity within a finite time as measured on the falling clock. In a Schwarzschild black hole the rising tidal forces would alternately stretch and squeeze the astronaut with ever-increasing frequency and intensity. In a spinning hole the stresses become infinite on a ring rather than at a point. Kerr immediately realised that this offered even more remarkable prospects. He excitedly told a colleague at the time, 'Pass through this magic ring and – presto! – you're in a completely different universe where radius and mass are negative!'

OUTSIDE THE HOLE

In the 1960s it wasn't clear whether the Schwarzschild and Kerr solutions describe a 'typical' black hole, or whether they were idealised and atypical holes which theorists had stumbled upon simply because they were the ones for which Einstein's equations could be most readily solved. We now know that there cannot be any other kinds of black hole. This realisation is crucial for our understanding of the 'real' universe. Viewed from outside, black holes are exactly standardised objects – no traces persist to distinguish how a particular hole formed, or what kind of objects it swallowed. The combined efforts of several theorists had proved this result by the early 1970s. Kerr's solution, when first discovered, seemed to describe a special and atypically symmetric situation. But it acquired paramount importance when theorists realised that it described space-time around *any* black hole. A collapsing object quickly settles down to a standardised stationary state characterised by just two numbers: one that measures its mass and one its spin. (In principle, electric charge is a third such number, but stars never have enough electric charge for this to be relevant to real collapse phenomena.)

Black holes are the 'ghosts' of dead massive stars; they have collapsed, cutting themselves off from the rest of our universe, but leaving a gravitational imprint frozen in the space they have left. Around black holes, space and time behave in highly 'non-intuitive' ways. For instance, time 'stands still' at the surface: an observer hovering there would witness the whole future of the external universe in what, subjectively, seemed quite a short period

To the physicist gravitational collapse is important, because (as described later) entirely new concepts are needed (just as in the initial instants of the big bang) to understand what is going on. The paradoxes within a black hole are as fundamental, and as far-reaching in their implications, as the puzzles that confronted Einstein's contemporaries at the beginning of the 20th century and triggered the development of relativity and the quantum theory. Black holes preclude space and time being a seamless continuum. They could even be gateways to other space-times sprouting from our own; the existence of black holes not only allows, but may even require, a broadening cosmic perspective that envisions our universe – everything astronomers might actually see – as just a member of an ensemble.

BLACK HOLES AS DEAD STARS?

But should black holes really exist? For more than 60 years, we've had good reasons for saying yes. In 1930, a precocious young Indian, Subrahmanyan Chandrasekhar, enrolled at Cambridge, where he hoped to become one of Eddington's students. During the long voyage to England, he thought about white dwarfs – the dense remnants of stars that can no longer draw on nuclear energy. He reached a startling conclusion. White dwarfs more than 1.4 times as heavy as the Sun couldn't exist: their central pressure could never build up high enough to counteract gravity.

This raised the question as to what happened when heavier stars ran out of fuel. They may, of course, throw off so much material in the course of their evolution that their masses end up below Chandrasekhar's limit for a white dwarf. The outer layers may be expelled in a supernova explosion, leaving a neutron star, as seems to have happened in the Crab Nebula. But there is a limit to how heavy a neutron star can be. This is less definite than the limit for a white dwarf, because it depends on how material behaves when squeezed to several times the density of an atomic nucleus, but is almost certainly below three solar masses.

Stars don't all have the 'prescience' to shed enough gas to bring them safely below this limit, and any stellar remnant more massive than about two or three solar masses would collapse completely when its nuclear energy ran out. The supernovae that are triggered by the heaviest stars, those above 20 solar masses, are thought to leave black holes rather than neutron stars; some massive stars may even collapse into black holes without producing any conspicuous supernova-type explosion. So the Galaxy should contain large numbers of gravitationally collapsed bodies. How can we find them?

THE X-RAY DISCOVERY OF BLACK HOLES

Once formed, black holes are essentially 'passive' and the best hope of locating them then lies in discerning their gravitational effects on neighbouring stars or gas. Among the first to think about this was Yakov Zeldovich. He has a prominent place in modern cosmology, not only through his individual contributions but because, from the 1960s onward, his research school in Moscow spearheaded so many key discoveries, even though cosmology and relativity had previously been ideologically tainted in the Soviet Union.

Zeldovich was one of the last polymaths of physics – in earlier phases of his career he had worked on particle physics, gas dynamics and other fields, despite having been preoccupied for many years with the Soviet H-bomb project. Zeldovich was trained as a technician, bypassing traditional higher education, and originally specialised in chemistry. He attributed his lack of early interest in physics to ultra-traditional schooling: his teacher read out Newton's laws first in Latin, and only afterwards in Russian.

Zeldovich realised that a black hole in close orbit around a star would accrete gas from its companion. Captured gas would swirl downwards towards the black hole, orbiting faster and faster as it got closer. Compression and viscous friction would make it so hot that it would radiate intense X-rays. If the inflow were turbulent and unsteady, the X-ray intensity would flicker irregularly.

X-ray observations highlight the most energetic features of the universe – the hottest gases, the strongest gravity, the most energetic explosions. As already mentioned, spinning *neutron stars* accreting from binary companions were among the first X-ray sources discovered. Cygnus X-1 is an X-ray source orbiting a binary companion; but it differs from the others, the accreting neutron stars, in that its X-rays vary erratically rather than with a regular period. Moreover, it weighs at least six solar masses, implying that it cannot be a neutron star or white dwarf. There are now several other black-hole candidates in our Galaxy at least as convincing as Cygnus X-1. These all have binary companions – ordinary stars orbiting around them, from which gas is being sucked inwards towards the hole. They are all too heavy[1] to be neutron stars; the rapid and irregular way their X-ray output varies fits with the black-hole hypothesis.

The case still isn't completely watertight: ad hoc interpretations that do not involve a black hole can always be devised. One's assessment of the odds obviously depends on whether one regards black holes as inherently absurd, or as plausible endpoints for stellar evolution. The alternative explanations are generally too contrived to be convincing: as the physicist Edwin Salpeter puts it, 'A black hole in Cyg X-1 is the most *conservative* hypothesis'.

THE LARGEST BLACK HOLES

The black holes I have discussed so far represent the final evolutionary state of stars, and have radii of 10–50 kilometres. But vastly larger black holes exist beyond our own Galaxy. The evidence for these, though more recent, is now even more compelling than for anything in our own Galaxy.

In the centres of some galaxies, gas and stars are swirling into a black hole weighing as much as millions, or even billions, of suns. They manifest themselves as quasars or as intense sources of cosmic radio emission. These holes, each as large as the Solar System, are regions that have already experienced the fate that will engulf everything if our universe eventually collapses.[2]

When its external supply of fuel ceases, a quasar switches off. The best guesses about quasar demography – their number, their lifetime and how many generations have lived and died – suggest that most galaxies went through a 'quasar phase', and may therefore contain 'relic' black holes, lurking in their centres.

Even a quiescent black hole, a dead quasar, still exerts a gravitational pull on its surroundings, and there would be two telltale signs of its presence in the centre of a galaxy. The first is simply a central 'blip' in the light from the galaxy, due to stars being pulled close to the hole. A second signature would be evidence that stars or gas clouds near the centre were moving anomalously fast.

Nearby galaxies obviously offer the best prospects for finding such effects. A giant galaxy called M 87 in the Virgo cluster of galaxies harbours a central dark mass of about 3 billion suns. This monster black hole would be bigger than the entire Solar System, Neptune and Pluto included. Another convincing case involves our nearest large neighbour in space, the Andromeda Galaxy. Its central hole weighs about 30 million solar masses. This evidence is strengthened by the Space Telescope, which yields sharper images than ground-based telescopes. Black holes exceeding a billion solar masses are so big that they could swallow a star in one piece; a star falling towards a somewhat smaller hole would be shredded apart first, creating a more conspicuous pyrotechnical display.[3]

The most convincing evidence for a supermassive black hole comes from the work of radio astronomers, who used a special technique (linking radio telescopes spread across an entire continent) to map the centre of a nearby galaxy, NGC 4258, in 100 times more detail than even the Space Telescope could achieve. They discovered a disc of gas orbiting exactly in the way expected if there is a central black hole of 36 million suns.

If there were indeed black holes in most nearby galaxies, our own Galaxy would seem underendowed if it didn't harbour one as well. The motions of stars and gas around the centre or 'hub' of our galaxy suggest a mass of 3 million suns – 10 times lower than Andromeda. Our own Galaxy can never have hosted a powerful quasar.

The plane of the Milky Way, in which the Sun lies, is pervaded with dust that obscures the view. It is therefore harder to be sure what is happening in our own Galactic Centre than in the centre of Andromeda. The leading authorities on this topic are Reinhard Genzel of Munich and Charles Townes of the University of California. In a 1988 article these authors assessed the net evidence for a black hole in our Galactic Centre as 'substantial but not fully convincing'. They later (1994) wrote another article, and I scanned it eagerly to find how their assessment might have changed. The evidence is now, they claim, 'holding up, and perhaps slowly growing, but by no means fully convincing'.

Galaxies often crash together and merge. If two merging galaxies each contained a massive black hole, then these holes would spiral towards each other in the centre of the merged galaxy, and coalesce. The coalescence of two black holes is a process involving nothing but the dynamics of space and time itself. Solving Einstein's equations in this violently unsteady, asymmetrical situation presents a challenge that even supercomputers cannot yet meet. During the coalescence a recoil could impart a powerful kick to the resultant (merged) hole, perhaps even ejecting it from its host galaxy. Some massive holes, created in galaxies, could now be hurtling through intergalactic space.

THE MATHEMATICAL PICTURE: SINGULAR MYSTERIES INSIDE THE HOLE

Einstein's equations can be readily solved only for especially simple cases – for instance, collapse of an exactly spherical object, or expansion of a 'model universe' that is completely uniform. Are these reliable guides to what happens in more realistic cases?

New mathematical concepts had to be deployed before theorists could analyse collapsing stars or expanding universes that are realistic rather than idealised. Roger Penrose, whose early training had been as a mathematician, was the catalytic figure. He introduced new mathematical techniques which revealed that 'singularities', where the strength of gravity 'goes to infinity', are deeply rooted in the structure of space and time. When the solution to an equation 'blows up' like this – when, as it were, 'smoke pours out of the computer' – it generally means that the theory has broken down, or become in some way inadequate.

Anything that implodes in an exactly symmetrical way obviously crashes together at a central point: the gravitational force then becomes infinite even in Newton's theory. This infinity is just an artefact of the symmetry – if the infall weren't exactly radial, the pieces would miss each other. But in Einstein's theory a *generic* collapse leads to a singularity – the extra kinetic energy of the transverse motions is equivalent to an extra mass, and therefore actually enhances the gravitational attraction pulling everything together.

Whenever a black hole forms, a singularity must develop inside it. Penrose first lectured on this idea in Cambridge in 1965. I was then a first-year graduate student, too inept mathematically to take in all he was saying. However, the implications were clear. Penrose had shown that Einstein's theory predicts its own incompleteness. 'Infinities' in a theory are a signal that some new physics intervenes. What happens is a mystery – space itself may change its nature on very tiny scales; extra dimensions perhaps appear; regions may 'pinch off' or even sprout into new universes. (These ideas reappear in Chapter 14.) Penrose's theorems also imply that there must have been a singularity at the beginning of our universe, even if the big bang were asymmetrical and irregular.

Penrose thinks not just more clearly than most of us, but in a very distinctive and unusual way. His geometrical grasp is unique. Mathematical physicists divide into two kinds. Some, having hit upon relevant equations, will thereafter think about the equations rather than the physical phenomenon itself (Chandrasekhar, for instance, thought very much in terms of equations). Others are more at ease with pictorial or geometrical concepts; Penrose is at the extreme 'geometrical' end of this spectrum – as adept at visualising four or more dimensions as most of us are in two dimensions.

He is remarkable in the breadth and originality of his insights. He spent many years developing his theory of 'twistors' – envisioning space and time as made up not of points but of interlaced 'light cones'. Even his recreational mathematics has been fruitful. He devised 'Penrose tiles' whose pattern never repeats, however large an area they cover. This work was taken up by crystallographers after so-called 'quasi-crystals' were discovered: these seem to have pentagonal symmetry, even though it was well known that no regular-repeating pentagonal lattice can exist. With his father Lionel (a geneticist) he devised the 'impossible objects' – perspective drawings that have no consistent interpretation as solid objects – which have become familar through being (for instance) the basis of Escher's engravings.

Many of these themes figured in Penrose's book *The Emperor's New Mind* – an eclectic romp throught all his enthusiasms. He addresses two of the most daunting fundamental problems: reconciling gravity with quantum mechanics; and the nature of consciousness and human thought and insight. His distinctive (and controversial) claim is that these two mysteries are linked. Even those unconvinced by the thesis can enjoy separate chapters on an *à la carte* basis. This book was an unusual best-seller. The sales pitch, 'Great scientist shows that human minds are more than "mere" machines', was a seductive one. But the uncompromising technicality of the book gave some baffled readers a nasty surprise.

In the early days of Einstein's theory, Eddington in Cambridge was its leading champion and expositor. Forty years later, when relativity underwent a renaissance, its most influential Cambridge exponent was Dennis Sciama (whose cosmological ideas have already featured in Chapter 2). It was Sciama who first enthused Penrose, originally a pure mathematician, to work on relativity. During the 1960s, Sciama attracted and inspired a steady flow of students, and thereby catalysed many of the key developments in relativity and cosmology. Among these students was Stephen Hawking. Sciama encouraged him to attend Penrose's lectures; these lectures expounded the mathematics that Penrose and Hawking utilised in their joint studies of gravitational collapse.

The results of this work were codified in a highly technical book *The Large-Scale Structure of Spacetime*, which Hawking wrote with George Ellis, another of Sciama's former students. Through this book, and his work on the nature of black holes, Hawking was, by the early 1970s, acknowledged as one of the leaders in relativity.

He was already physically frail. None of us then predicted the astonishing later phases of his career. His most remarkable single discovery, black-hole evaporation, came in 1974 (and is described in Chapter 11). But that was itself just the impetus for a crescendo of achievement that continues to this day. Nobody else since Einstein (except perhaps Penrose) has contributed more to our understanding of gravity. And no physicist since Einstein has achieved such worldwide fame.

When Hawking received an honorary degree from Cambridge, the Orator quoted the encomium of Epicurus by Lucretius: 'The living force of his mind overcame and passed far beyond the flaming ramparts of the universe, traversing in mind and spirit the boundless whole.' Admiration for his sustained achievements has been hugely amplified by the stark juxtaposition of his physical constraints and his mind 'roaming the cosmos'. Public fascination would be more muted if he had similar standing in (for instance) cell biology or chemistry.

Hawking's public celebrity stemmed, of course, from a book very different in style from the one he wrote with George Ellis – his popular *A Brief History of Time*. The marvellous feature of that book was that it should have been written at all. After finishing a rough draft, Hawking suffered a further setback to his health, which left him, for a time, completely immobile and unable to communicate except by directing his eyes towards the appropriate part of a large board on which the alphabet was written. Without computer technology, he would never have been able to convey any thoughts beyond the most basic requests. But a lever-controlled word-processor enabled him, albeit slowly and painfully, to complete the book. A speech-synthesiser enabled him to converse more clearly than before, and even (with careful preprogramming) to present hugely popular public lectures. Advances in machine translation techniques may soon allow him, without extra effort, to address audiences in Japan and Korea in their own language.

Scientists generally limit their writing to technical papers that appear in academic journals after being 'refereed' by a (usually anonymous) colleague who is supposed to assess whether they are worthy of publication This is the formal system whereby the consensus builds up. But these journals – what scientists call 'the literature' – are impenetrable to non-specialists. They now exist just for archival purposes, largely unread even by researchers, who depend more on informal 'preprints', electronic mail and conferences. A colleague once told me that the average number of readers of a scientific paper was 0.6 (and went on to wonder, more cynically, whether this included the referee). The essential *content* of these papers – or at least of those that survive an intellectual sifting – eventually, of course, diffuses to a wider public. Most of us would derive less satisfaction if our research never percolated beyond other specialists. Stephen Hawking's impact and outreach, through *A Brief History of Time*, has been unparalleled. This success had one negative consequence: the book came to the attention of philosophers and theologians, and received more scrutiny than it could really bear.

WAS EINSTEIN RIGHT ABOUT GRAVITY?

The renaissance in gravitational research that began in the 1960s was due partly to more powerful mathematical techniques; it was also stimulated by observational discoveries. For the first time, astronomers realised that there were places in the universe – even within our own Galaxy – where relativistic effects could have extraordinary implications. The exhilaration at that time was expressed by Thomas Gold. In an after-dinner speech at the first big conference on the new subject of 'relativistic astrophysics', held in Dallas in 1963, he said 'the relativists with their sophisticated work [are] not only magnificient cultural ornaments but might actually be useful to science! Everyone is pleased: the relativists who feel they are being appreciated, who are suddenly experts in a field they hardly knew existed; the astrophysicists for having enlarged their domain . . . it is all very pleasing, so let us hope it is right.'

General relativity has the virtue of being highly specific. Any single discrepant observation or experiment would be deadly – the theory could not be brought into line by any minor tweaking or small adjustments. This is the motivation for continuing to devise new tests. The discovery of black holes opens the way to testing the most remarkable consequences of Einstein's theory.

Objects like Cygnus X-1, and the centres of galaxies, are places where the space of our universe gets punctured by the accumulation and collapse of large masses – collapse to entities described exactly by fairly simple formulae (the 'Kerr solution' to Einstein's equations).

As Roger Penrose has remarked, 'Is it ironic that the astrophysical object which is strangest and least familiar, the black hole, should be the one for which our theoretical picture is most complete.'

The very massive holes offer the best prospects for confronting observations with gravitation theory. A black hole of stellar mass develops only after collapse to nuclear densities, with all the physical uncertainties entailed by high-density physics. In contrast, the black holes in galactic nuclei, some as massive as billions of suns, could have formed without being denser than air.[3]

The radiation from such objects comes from hot gas swirling downward into a deep 'gravitational pit'; it displays huge Doppler effects, as well as having an extra redshift because of the strong gravity. Measurements of this radiation, expecially the X-rays, can probe the flow very close to the hole, and diagnose whether the 'shape of space' near it agrees with what Schwarzschild and Kerr predict.[4]

Still deeper mysteries lie *within* the black hole. Conditions at the 'surfaces' of huge black holes like the one in M 87 are not inhospitable. A star could fall within the horizon before being destroyed; so (in principle) could an astronaut. There would then still be several hours, or even days, for leisured observation before being discomforted by approach to the central singularity. But what happens then? If the hole was not rotating, and had been undisturbed for a long period, then the tidal forces would stretch an infalling astronaut radially. But, in a more realistic hole, the tidal forces would shake him with increasing violence – 'spaghettification' is the colloquial term for this fate. More speculatively, there could be an eruption into a new space, perhaps even into a new inflationary universe (see Chapter 14). This is the relevance of black holes – these whirlpools in the centres of galaxies, and their small-scale counterparts orbiting companion stars in our own Galaxy – to the broader cosmological perspective.

COHERENT AND INCOHERENT PROGRESS

General relativity, or 'gravitational physics', forms an interesting case history for students of the history of science. The key conceptual advances since the 1960s can be traced to collaborations and interactions between a small number of leading workers. These people nearly all emerged from three research 'schools' – those led by Zeldovich in Moscow, Sciama in Cambridge and Wheeler in Princeton. Moreover, the interactions among them were almost universally cooperative and constructive (A recent book by the American theorist Kip Thorne gives an individualistic perspective on this research community.) In these respects it is atypical of the advances that feature in this book: science normally progresses in more boisterous and less coherent ways.

Astronomers are explorers: serendipity still plays a big role in what they do. Few phenomena have been successfully predicted, though theorists often feel (and even sometimes say), 'With hindsight, I could have predicted that.' Most discoveries have surprised theorists, and initially perplexed them. Sometimes the picture quickly clarifies: a plausible interpretation emerges, and a consensus is quickly established. Sometimes we fruitlessly try to interpret fragmentary data, when only a year or two later further evidence reveals that all the prior ideas, or (if we're lucky) all but one, were non-starters. The microwave background, for instance, was quickly recognised as a relic of the early universe (Chapter 3); nor did it take long before pulsars (Chapter 4) were accepted to be spinning neutron stars.

Not all phenomena quickly fall into place: sometimes the mystery takes decades to dispel. When one looks back, for instance, at quasar research over the 30 years since these superluminous objects were discovered (see Chapter 2), progress seems depressingly slow. Sometimes we had the illusion that the subject was advancing rapidly, but we have really had a rather slow advance, with 'sawtooth' variations superimposed on it as fashions came and went.

Quasars were, in a sense, discovered too early. If they had been discovered after the theory of black holes had been fully worked out, and after pulsars and compact X-ray sources had familiarised us with the likely efficiency of gravitational power sources, a consensus would have developed more quickly about what their 'central engine' really was.

Many of the bizarre ideas current in the early days of quasar research have now been abandoned. But a clean-cut refutation happens only rarely in astrophysics: rival interpretations persist for a long time. A cynic might argue that they often survive only because their proponents are adept at replacing or patching up faulty parts to keep shaky old vehicles 'roadworthy'. Such an attitude is not necessarily justified, and, to explain why, I must digress briefly into methodology.

The way we are told science is done is like this: the data suggest a hypothesis, which in turn suggests further tests, whereby the original hypothesis is either refuted or refined. This simple procedure is realistic in, for instance, particle physics, where the fundamental entities may be exactly reducible to a few basic constants and equations. But other sciences deal with inherently complex entities and no theoretical scheme can be expected to account for every detail. In geophysics, for instance, the concepts of continental drift and plate tectonics have undoubtedly led to key advances; but we shouldn't expect them to explain the precise shapes of America and Africa. Attempts to understand cosmic phenomena should focus on those features of the data that genuinely test crucial ideas, and not be diverted into trying to interpret something that's accidental or secondary.

All too often, our picture is a kind of rough caricature; though we may hope that, like a good caricature, it highlights rather than obscures the essence of the phenomenon.

EPILOGUE

After his pioneering student insights into how stars end their lives, Chandrasekhar (universally known as 'Chandra') shifted to other topics. His style of research was unusual. He would choose a subject and explore it thoroughly for a few years. He would then systematise his thoughts into a book, and move on to something else. He produced classic texts on stellar structure, dynamics of stellar systems, fluid mechanics and other specialised topics. But he returned to the study of black holes in much later life.

The early 1970s were the 'heroic age' of black-hole research. Theorists discovered that, if Einstein was right, black holes weren't infinitely diverse but standardised objects, characterised just as surely as any elementary particle by their mass and their spin. And astronomers were starting to suspect that black holes were not just theoretical constructs, but might actually exist in our universe.

This made a deep impression on Chandra aesthetically as well as scientifically. In a lecture in 1975 he said:

> In my entire scientific life . . . the most shattering experience has been the realisation that an exact solution of Einstein's equations of general relativity, discovered by the New Zealand mathematician Roy Kerr, provides the absolutely exact representation of untold numbers of massive black holes that populate the Universe. This 'shuddering before the beautiful', this incredible fact that a discovery motivated by a search after the beautiful in mathematics should find its exact replica in Nature, persuades me to say that beauty is that to which the human mind responds at its deepest and most profound.

Chandra was already in his sixties when he embarked on black-hole research. He was fond of quoting the great physicist Lord Rayleigh's response to T.H. Huxley's claim that 'scientists over 60 do more harm than good'. Rayleigh (aged 67 at the time) had responded, 'That may be, if he undertakes to criticise the work of younger men, but I do not see why it need be so if he sticks to the things he is conversant with.' It is a precept that Chandra himself manifestly followed. He never fully absorbed the mathematical techniques introduced by Roger Penrose, which had given the subject such impetus. Instead, he made his distinctive contribution by adapting the more 'classical' methods he had used in other contexts.

He analysed how black holes would respond when their equilibrium was perturbed, extending techniques that had traditionally been used to study the vibration modes of a drum, or of earth and oceans. The techniques, in a sense, complement the methods Penrose pioneered: they cannot handle 'generic' collapse, where there is no special degree of symmetry; but they yield a more quantitative picture of what would happen if a black hole were perturbed (by, for instance, a smaller object falling into it or orbiting close to it). These techniques offer a 'probe' for black holes, just as seismologists can learn about the Earth's structure from the various modes of oscillation when its crust is set 'ringing' after an earthquake.

Chandra was unique among his contemporaries in his intellectual stamina, which, combined with his self-disciplined neatness, allowed him to carry through the most elaborate mathematical manipulations without flagging and (equally remarkably) without mistakes. I recall the first time I heard him lecture, at a Cambridge seminar. He presented his mathematics on slides, which he ran through at bewildering speed because each equation was too long to fit on a single slide, and spilled over on to several. He ended his talk with a typical disclaimer: 'You may think I have used a hammer to crack eggs, but I have cracked eggs.'

Chandra's mathematical virtuosity is dauntingly manifest in his 650-page treatise, *The Mathematical Theory of Black Holes*. In one chapter, 100 pages long, the manipulations are so heavy and the argument so terse that he adds the following footnote:

> The reductions that are necessary to go from one step to another [in this chapter] are often very elaborate and, on occasion, may require as many as ten, twenty, or even fifty pages. In the event that some reader may wish to undertake a careful scrutiny of the entire development, the author's derivations (in some 600 legal-sized pages and in six additional notebooks) have been deposited in the Joseph Regenstein Library of the University of Chicago.

Such is the aura surrounding Chandra and his subject that this formidable and recondite text has notched up several thousand paperback sales. Its sales are not, of course, in the class of Hawking's *A Brief History of Time*, but it has probably surpassed Hawking's book in the ratio of copies sold to copies actually read. Any reader who perseveres would echo the 19th-century scholar William Whewell's reaction to the mathematics of Newton's *Principia Mathematica*: 'We feel as when we are in an ancient armoury where the weapons are of gigantic size; . . . we marvel what manner of men they were who could use as weapons what we can scarcely lift as a burden.'

Chandra was 72 when his book on black holes appeared. Most of us suspected that it would be his final monograph – indeed, it rounded off his career with a fitting symmetry, by codifying our understanding of the objects foreshadowed by what he had done as a student in Cambridge, decades before (an initial insight that gained him as much acclaim as the ensuing 50 years of intellectual toil). But he continued his unflagging output of highly technical papers.

And he developed a new enthusiasm. His lifelong fascination with individuals who scale the supreme peaks of creativity, whether in science or the arts, led him to a detailed study of Newton's work, which culminated in a 600-page exposition, *Newton's Principia for the Common Reader*, published in 1995.

Chandra intended his book on the *Principia* to be his final one. Indeed he had decided, at the age of 84, to bring his extraordinarily sustained scientific efforts to a close. He was always critical of elderly scientists who lived on their reputation; a clean break was better than the risk of compromising his standards. As he told his colleagues, 'There is a time for all things, and a time to end all things.' He died in August 1995; his last paper appeared that same month. Whether he would ever actually have relaxed his arduous regime, or made a clean break from a long life of thinking, learning and working, we'll never really know.

As they grow older, some scientists cease doing research. Others retain their urge to understand the world, but no longer derive satisfaction from 'routine' problems; such people overreach themselves, often embarrassingly,[5] by tackling (and even claiming to solve) fundamental problems beyond their real expertise. Neither of these paths tempted Chandra.

NOTES

1 Astronomers can measure the Doppler effect in the light from the companion star, and infer how fast it is orbiting. Newton's laws of gravity then tell us how heavy the X-ray source itself is (just as we can infer the Sun's mass by knowing how fast the Earth is moving, and the size of its orbit). The sources that are thought to be neutron stars all have masses close to 1.4 solar masses; the black-hole candidates (those that flicker irregularly) are several times heavier.

2 Massive black holes can generate power in two ways. If they accrete gas or even entire stars from their surroundings, the captured material swirling down into them can, before being irrevocably swallowed, convert about 10 per cent of its rest-mass energy (mc^2) into radiation. This is a scaled-up version of what may be happening in Cygnus X-1. But a second, and more interesting, process was discovered by two Cambridge colleagues, Roger Blandford and Roman Znajek. Black holes behave like gyroscopes or flywheels. Blandford and Znajek showed that a magnetic field from outside, which may come from gas or stars in the host galaxy, can 'apply the brakes' to a spinning hole, and thereby tap its latent spin energy. The strong cosmic radio sources (Chapter 2) are probably energised in this way. Astrophysicists aim to calculate how much power is derived from accretion, and how much is extracted from the hole's spin, and to find the form in which these respective contributions emerge. Such calculations play the same part in the modelling of activity in galactic centres that nuclear physics does in theories of stellar structure and evolution. The phenomena in galactic centres have been one of my main research interests, but are peripheral to the theme of the present book.

3 The radius of a hole scales with its mass; its volume therefore goes as

the cube of its mass. The density an object has to attain before closing off from the external universe and becoming a black hole therefore scales as mass/volume, or as $1/(\text{mass})^2$.

There is a limit to how closely a star can approach a black hole without suffering damage. Such a star would be sensitive to tidal effects – the *gradient in* the gravitational pull across the star. These forces, if too strong, would tear the star apart. The tidal forces at their 'surfaces' are more gentle for the biggest holes; the one in M 87 could swallow a solar-type star without disrupting it. However, if the hole's mass is below 100 million suns (the range relevant to the nearest galaxies) then a star like the Sun would be tidally disrupted if it came within 10 times the hole's radius. (A very compact star – a white dwarf, for instance – could fall inside a massive black hole more or less intact.) A star responds in a complicated way, being stretched along the orbital direction, squeezed at right angles to the orbit, and strongly shocked. This phenomenon poses an as-yet-unmet challenge to computer simulations. Within a few years, detailed computer modelling should allow us to calculate the characteristics (duration, colour and so on) of the flares that occur when a star is disrupted. Astronomers can then search for evidence of these disruptive events, which will provide direct evidence on conditions very close to the hole.

4 A fuller account of the more observational topics mentioned in this and the previous chapter is given in the book *Gravity, Fatal Attraction: the Discovery of Black Holes*, by Mitchell Begelman and myself (W. H. Freeman, 1996).

5 Arthur Eddington, from 1930 onwards, immersed himself in a numerological 'fundamental theory' that struck little resonance with his contemporaries (see page 251). After one of his lectures, a concerned student in the audience asked his adviser, the physicist Samuel Goudschmidt, how to avoid going the same way in later life. Goudschmidt reassured him: 'Don't worry. Only geniuses get like that; the rest of us just get dumber and dumber.'

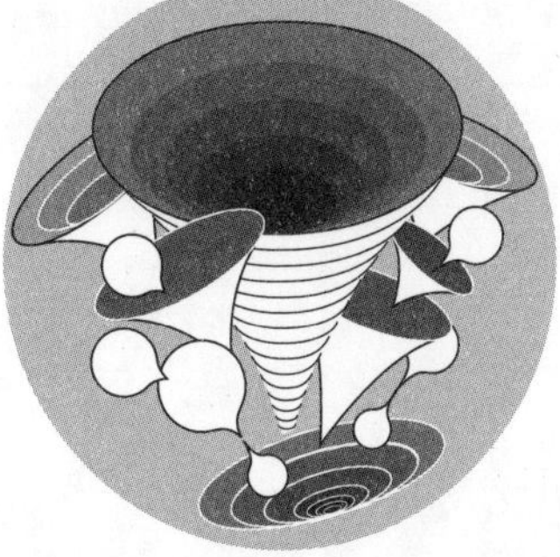

6

IMAGE AND SUBSTANCE: GALAXIES AND DARK MATTER

> As the visible Creation is supposed to be full of Siderial Systems and Planetary Worlds, so on, in like similar Manner, the endless immensity is an unlimited Plenum of Creations, not unlike the known Universe.
>
> Thomas Wright of Durham (1752)

ONE OF THE MOST STRIKING images from space shows the Earth at night. At first glance there seems no pattern. But one then picks out some conspicuous features – light from major cities, burning oil wells in the Middle East, and the glow from millions of wood-burning stoves in Indian conurbations – and thereby discerns the familiar underlying pattern of the continents and their coastlines. But most things on Earth don't shine, and, were this our only picture of the Earth, any inferences would be biased and incomplete.

So it is when we look outward at the cosmos. Optical telescopes remain pivotal, and still provide more information than any other technique: stars radiate most of their energy as visible light, and the Earth's atmosphere is transparent to this radiation. Radio astronomy, however, offers a distinctive window on universe. The most prominent objects in the radio sky are quite different from those that dominate in optical photographs.

It is no coincidence that our eyes have evolved to be sensitive to the dominant radiation from the Sun. But there are many other kinds of radiation (ultraviolet and X-rays, for instance) to which our eyes can't respond and to which the Earth's atmosphere is opaque. The emission from exotic cosmic objects spills over into these other bands. For instance, X-ray observations from spaceprobes (see Chapter 5) were crucial in the quest for black holes.

Our perception of even the most familiar galaxies is changing dramatically. They are ten times bigger and heavier that we used to think. The entities that conventional astronomers observe and call 'galaxies' are no more than traces of sediment trapped in the centres of vast swarms of invisible objects of quite unknown nature. The gravity of this dark matter holds galaxies together and moulds their structures.

This new perspective has emerged because a variety of new techniques complement the efforts of traditional astronomers. Observations from spacecraft, and even delicate experiments at the bottom of deep mineshafts, are shaping our new conception of what galaxies really are.

OUR GALAXY AND OTHERS

It is over 400 years since Copernicus dethroned the Earth from the privileged position that Ptolemy's cosmology accorded it, and described the general layout of the Solar System in the form accepted today. But the realisation that the Sun's position wasn't special came about very gradually and is still progressing; our perception of the scale of the cosmos is still enlarging. In the 18th century William Herschel interpreted the Milky Way as a flat disc-shaped structure of stars in which the Sun was embedded. The great philosopher

Immanuel Kant argued at that time that some 'nebulae' were separate systems, rather than being part of our Milky Way, but only after 1920 was our Galaxy (the flat disc that we see as the Milky Way) realised to be just one fairly typical galaxy similar to the hundreds of millions of others that can be registered by a large telescope, and are the basic units making up our large-scale universe.

The apparent faintness of the stars (other 'Suns') in our night sky tells us how exceedingly dispersed they are relative to their actual physical sizes. (If the Sun were scaled down to a sugar-lump, the nearest stars would be a thousand miles away.) There is therefore little risk that another star will ever crash into our Sun, or even come close enough to dislodge planets from our Solar System.

Our Sun is held in orbit around the Galactic Centre by the inward pull of gravity, just as the planets do in orbiting the Sun. It moves at 250 km/sec, making a circuit in about 200 million years (a 'galactic year'). If our own Milky Way were viewed by a distant observer, the Sun would lie about two-thirds of the way out towards the edge of the visible disc.

THE ZOO OF GALAXIES

Galaxies are to astronomy what ecosystems are to biology. Each galaxy undergoes complex internal evolution. Single stars, the individual organisms in a galactic ecosystem, can be traced from their birth in gas clouds, through their life cycle, to their (sometimes explosive) deaths. The atoms we are made of come from all over our Milky Way Galaxy, but few come from other galaxies.

Clouds of interstellar gas are even now condensing into new stars; the Space Telescope has given us spectacular pictures of the Eagle Nebula, and other clouds of gas and dust where this is happening. Luminous blue stars (like, for instance, the famous 'Trapezium' stars in Orion), burn their nuclear fuel so fast that their lifetimes are relatively short, and testify that star formation is a continuing process. When these stars die, they will return much of their material to the interstellar gas. The basic process going on in our Galaxy is, in effect, a 'cycle': gas condenses into stars; some subsequently rejoins the diffuse interstellar medium via stellar winds and supernova explosions and is available for forming a new generation of stars.

Every atom of carbon, nitrogen and oxygen in the Solar System was synthesised in early stars which died before the Sun formed. Some of the material that turns into stars is, however, permanently trapped, in the sense that it gets incorporated into low-mass long-lived stars or into the compact remnants left behind when a star becomes a supernova.

Just as we are at a typical place within our Galaxy, our Galaxy is itself typical of a common type.[1] Most galaxies can be characterised as either 'disc' or 'elliptical' systems; in the latter, the stars are not confined to a disc, but swarm around in chaotic orbits, each star feeling the gravitational pull of all the others.[2]

The 'rate of metabolism' is not the same in all galaxies. The conspicuous spiral arms in some disc galaxies delineate regions where young bright stars are forming unusually rapidly. They correspond to some kind of persistent wave pattern in the disc, but there is still no completely satisfactory explanation of what excites and maintains such a wave. Ellipticals used up most of their gas long ago, and few new stars are still forming. Disc galaxies have not yet got so close to the final state in which essentially all the gas is tied up in low-mass stars or dead remnants.

Some galaxies have been distorted by the gravitational pull of a very close neighbour: some are even colliding and merging with a close companion. If we could handle a galaxy in the laboratory we would probe and perturb it in various ways, and see how it responded: nature performs just such 'experiments'. Computers can, of course, simulate encounters between galaxies with growing realism, but these interacting galaxies allow us to compare such simulations with the real thing.

We have learned how stars form, why they shine, and how they evolve. Why *galaxies* themselves exist is less straightforward than the equivalent question for stars. Galaxies formed at an earlier and remote cosmic epoch (see Chapter 7). We don't know which features can be explained in terms of ordinary processes accessible to study now, and which have to be traced back to causes in the earliest universe – galaxies are affected by 'genetics' and by environment.

But a deeper mystery must be addressed first. To our embarrassment, 90 per cent of every galaxy is unaccounted for; what we actually see is no more than 10 per cent of their total substance. All the rest is in some mysterious 'dark' form. Obviously, we won't get far without knowing more about this dominant constituent.

THE QUEST FOR DARK MATTER

Nineteenth-century astronomers noted that the planet Uranus was straying from its predicted orbit. Urbain Leverrier in Paris and John Couch Adams in Cambridge suspected that the gravitational pull of another planet was causing these deviations, and used Newton's laws to calculate where in the sky that planet should be. Neptune was thereby discovered in 1846 by Johannes Galle at the Berlin Observatory – not, sadly, by Adams's compatriots, whose lethargy is an inglorious episode in British astronomy. (A 12-inch refracting telescope whose main claim to fame is that it failed to discover Neptune survives in Cambridge.)

The same basic technique is now used to infer the presence of unseen planets orbiting other stars (see Chapter 1), and to 'weigh' black holes by monitoring the motion of stars orbiting close to them (see Chapter 5). Applied on a grander scale, similar methods have transformed our perception of what our universe is made of. Everything astronomers observe turns out to be a small and atypical fraction of what exists.

The discs of our Galaxy and Andromeda contain clouds of gas, as well as stars. This gas is mainly hydrogen. Atoms of hydrogen emit radiation at a distinctive wavelength in the radio band. Radio astronomers can therefore detect these clouds and infer (from the Doppler effect) how fast they are moving. Some clouds are a very long way out – their orbits lie far beyond the outer limit of the optically detectable disc. If the outermost gas were feeling just the gravitational pull of what we can see, it should move more slowly, just as Neptune and Pluto orbit the Sun more slowly than the Earth does. But the far-out gas is moving just as fast as the gas closer in. This tells us that a heavy invisible halo surrounds our galaxy – just as, if Pluto were moving as fast as the Earth, we should have to infer a heavy invisible shell outside the Earth's orbit but inside Pluto's.

On still larger scales – entire clusters of galaxies, each several million light-years across – the message is the same. The argument here dates back 60 years, to Fritz Zwicky, the Swiss-American physicist whose speculations about supernovae and neutron stars were noted in Chapter 4.

The random motions of its constituent galaxies tend to disperse a cluster; this disruptive tendency is balanced by the effect of gravity, which, if the galaxies had no relative motions, would cause them to fall together at the cluster centre. The speeds of the galaxies (or at least the part that is directed along our line of sight) can be measured from the Doppler effect. Zwicky was perplexed to find that the galaxies were moving so fast that clusters should be flying apart – to hold clusters together required the gravitational pull of something far heavier than the galaxies themselves.[3]

Zwicky actually had another idea for weighing clusters, that only in the 1990s has been seriously implemented: measuring how strongly they deflected light. (Bending of light rays from distant stars by the Sun's gravity, observed during a total eclipse, offered a famous early confirmation of Einstein's general relativity.) Pictures of clusters, especially those taken with the Space Telescope, are sharp enough to reveal large numbers of very faint galaxies lying far beyond the cluster. Some of these appear distorted into long streaks or arcs, because gravity makes the cluster act like a huge lens. We can infer the amount of dark matter, and how it is spread through the cluster, from the precise way the images of background objects are distorted and magnified. These huge natural lenses offers a bonus to astronomers interested in how galaxies evolve, because they bring into view very remote galaxies that would otherwise be too faint to be seen.

Without dark matter, clusters would fly apart – indeed, they could never have formed in the first place. Dark matter is the dominant gravitational influence on the cosmos – all the large-scale motions of galaxies are induced by (or responses to) its gravitational pull. This realisation alters our perspective on our universe. But we really have no reason to be surprised – there are all too many forms dark matter could take, and the aim of observers and theorists must be to narrow down the range of options.

WHAT CAN THE DARK MATTER BE?

In his later years, Zwicky propagandised what he called the 'morphological method', a systematic procedure to explore all conceivable possibilities. Without such an aid to creative thinking, we tend, through limited imagination, to overlook many options. A 'morphological' approach is certainly needed when we are faced with the baffling problem of dark matter. The 'obvious' candidates certainly do not exhaust all options.

Small faint stars are the usual suspects for dark matter in galaxies. Stars below eight per cent of the Sun's mass wouldn't be squeezed hot enough to ignite the nuclear fuel that keeps ordinary stars shining. They're called 'brown dwarfs' – a term coined by Jill Tarter, the American astronomer who was among the first to theorise about them. How many might we expect? Theory can so far tell us little. The proportions of big and small stars that form are determined by processes within interstellar clouds as complicated as those that govern the climate here on Earth. Not even the most powerful computers can tell us what happens: the processes are intractable, for the same reasons that weather prediction is difficult.

A huge population of brown dwarfs could have formed when the Galaxy first condensed from primordial gas. Perhaps there are many objects with even lower masses, more akin to planets than to stars.

Could the dark matter be black holes or neutron stars – remnants, perhaps, of earlier generations of heavy stars which all died long ago? This is one option that can actually be dismissed – and shortening the candidates' list is progress of a kind. The precursor star of each such remnant would, during its bright active life, have made carbon, oxygen and other elements further up the periodic table. This 'processed' material would be ejected, via stellar winds or supernova explosions. If there were enough remnants to constitute the dark matter in our Galaxy, then their precursors would have produced far more carbon, oxygen and iron than we find.

There is one escape from this conclusion – perhaps these heavy atoms all *fall into* remnant black holes rather than being flung out in an explosion. That could happen only if the precursor stars weighed hundreds of times more than the Sun. Such ultra-massive stars would never explode. Instead, when they have exhausted their nuclear fuel the central pressure suddenly drops, and they implode into a black hole which swallows all the processed material. No stars as heavy as this are now seen forming, anywhere. But the Galaxy could have been lit up by a generation of ultra-massive stars in the remote past, when it was young; their remnants could then constitute the dark matter.

So the dark matter could be in heavy black holes or in brown dwarfs. These are the most straightforward options, which involve only a modest extrapolation (upwards or downwards in mass) from familiar stars. But, if we're prepared to extrapolate further, there are other options – small rocklike bodies, lumps of frozen hydrogen, for instance. Zwicky himself (following his 'morphological method') speculated about 'nuclear gremlins' – small lumps of matter as dense as a neutron star. Edward Witten, the great mathematical physicist from Princeton, advanced a similar speculation. He proposed quark nuggets – lumps in an exotic dense phase, surviving from the very early universe.

SEARCHES FOR LENSING

Heavy objects, even dark ones, bend light rays passing close to them. They can act like lenses, magnifying more distant stars by focusing their light towards us. If a dark body (a brown dwarf or a black hole, for instance) moves across the line of sight to a background star, the magnification rises, reaching a peak when the alignment is closest, and then declines again; the background star consequently brightens and fades in a predictable way. The required alignment is very exact, so this phenomenon shouldn't happen very often: even if there were enough brown dwarfs to make up all the dark matter in our Galaxy (and there would need to be several trillion!), the chance that any particular background star is magnified is less than one in a million.

To stand a chance of detecting a magnified stellar image one must either be prepared to wait for a very long time indeed or (more optimistically) raise the odds by observing not one but millions of background stars. Until recently, this task seemed too daunting. But in 1993 a group of US and Australian scientists rejuvenated a derelict 100-year-old telescope, equipping it with the latest light-detectors and computer controls; French scientists used a small telescope in Chile for a similar project. Both of these groups monitored, every clear night, several million stars in the small nearby galaxy (about 150 thousand light-years away) known as the Large Magellanic Cloud. And Polish astronomers, using a small observatory in Chile, have looked for lensing events towards the centre of our own Galaxy, where the stellar concentration is high.

Many astronomers make their living out of studying pulsating stars, flare stars and binaries; these searches yield a rich harvest of such stars. But for those seeking evidence of gravitational focusing, *intrinsically* variable stars are a nuisance. The challenge is to pick out rare instances of *apparent* variability characteristic of a lensing event – a symmetrical rise and fall, with no change in colour.

Several convincing lensing events have already been discovered, just of the kind that would be caused by small faint stars in our Galactic Halo. However, if there were enough to make up the entire Halo mass, there should have been at least twice as many events as have been seen. So we are still left seeking another candidate for the dark matter. (It will be harder, however, to discover, or exclude, heavier black holes. These would cause similar lensing events, but more rarely and with a much slower rise and fall.)

The scientists who spearheaded these searches for lensing were, incidentally, people with little observational experience. Either their expertise lay in particle physics, or they were theoretical astrophysicists. There was nothing special about the telescopes or instruments needed, but the sheer amount of data daunted 'traditional' astronomers. Physicists accustomed to experiments in particle accelerators, where millions of collision events are recorded – only a few of which are interesting – were less easily discouraged; likewise the theorists, who perhaps didn't appreciate the practical difficulties. Whatever the eventual outcome of the quest, the optimists are already vindicated. It is indeed feasible to monitor millions of stars, and use computers to pick out the occasional one that varies in a distinctive way.

RELIC PARTICLES FROM THE FIREBALL AS DARK MATTER?

It would, in a way, be disappointing if all the dark matter turned out to be low-mass stars or black holes. Physicists would certainly be more excited if exotic particles were involved. Neutrinos are one option. These particles are made inside very hot stars, and in the early universe. Indeed, neutrinos left over from the 'cosmic fireball' should be almost as abundant as photons: there would be hundreds of millions of them for every atom in the universe. Because neutrinos so greatly outnumber atoms, they could be the dominant gravitating stuff even if each weighed only a hundred millionth as much as an atom. Before the 1980s, however, almost everyone believed neutrinos were 'zero rest-mass' particles; they would then move at the speed of light, and carry energy, but their gravitational effects would be unimportant. (Likewise, the photons left over from the early universe, now detected as the microwave background radiation, don't now exert any significant gravitational effects.)

In 1979 Valentin Lyubimov in Moscow claimed to have measured a neutrino's mass. His experiment was never repeated, and most physicists now discount it. But it stimulated cosmologists to take non-zero masses more seriously. A further claim in 1995, by a research group in Los Alamos (using different techniques), was also controversial. It appeared in print, in a paper signed by 39 authors; but one dissident member of the group published, in the same issue of the journal, a separate analysis of the data reaching an opposite conclusion! Observations or experiments, even when they turn out to be wrong, often still provide a positive impetus to science, simply because they stimulate theorists to explore novel possibilities that might otherwise be overlooked.

The nearby supernova observed in 1987 (see Chapter 1) also offered some clues about neutrino masses. The sudden collapse of a star's core that triggers a supernova (and which leaves a neutron star, or perhaps a black hole, as a remnant) releases a colossal pulse of energy, which escapes mainly as neutrinos. There are about 10^{57} atoms in the collapsing core; several neutrinos are created for each atom, so the supernova would generate about 10^{58} of them altogether. Ordinary atoms are almost 'transparent' to neutrinos; nearly all the neutrinos that hit the Earth would have gone straight through. A few, however, were caught by very sensitive instruments. A Japanese experiment (Kamiokande), deep underground in a salt mine, recorded 11 events; an American experiment (in a zinc mine in Ohio) recorded eight more. These numbers pleased astrophysicists, because they fitted well with what supernova theories predicted.

These experiments also tell us something about neutrino masses. If these masses weren't zero, neutrinos from a supernova would move nearly as fast as light, but not quite. Those detected in

1987, after travelling 170 000 light years from the supernova, all arrived within a few seconds of each other. This ruled out a mass as high as Lyubimov had claimed. But neutrinos are still in the running as dark-matter candidates: another type, called the 'tau' neutrino, should be heavier than those coming from supernovae. So the dark matter could still be tau neutrinos surviving from the early universe.

Neutrinos at least exist. But particle theorists have a long shopping list of particles that *might* exist, and (if so) could have survived from the early phases of the big bang. These hypothetical particles, heavy but electrically neutral, would generally, like neutrinos, go straight through the Earth. A tiny proportion, however, interact with an atom in the material they pass through, releasing a minuscule amount of heat or sound which sensitive experiments could measure. To detect these rare events – maybe one per day within every kilogram of material – experimenters must go deep underground, to reduce the 'background' from other kinds of event that would confuse or drown out the signal being sought. Several groups of physicists have taken up the challenge.[4]

Only an extreme optimist would bet more than evens that these dedicated experimenters down their mine shafts will detect anything. But the goal is still worth shooting for. A positive result would not only reveal a new class of elementary particle which could never be made in terrestrial accelerators, but also tell us what 90 per cent of our universe was made of – success would be at least as momentous as Penzias and Wilson's discovery of the microwave background in the 1960s.

Dark-matter theories are no longer unconstrained. Serious searches are under way for the various candidates. Gravitational lensing may detect enough faint stars or black holes; underground experiments may reveal some new kind of particle which pervades our Galactic Halo, or at least set limits on the tenable options. It should quench hope of quick progress when we note that our universe is mainly made of entities whose individual mass may range anywhere from 10^{-33} gm (exotic particles) up to 10^{39} gm (heavy black holes) – an uncertainty of more than 70 powers of 10. Astrophysics may not always be an exact science, but seldom is the uncertainty as gross as this.

It is not, however, wishful thinking to expect more than one important kind of dark matter – maybe exotic particles pervade large clusters and superclusters, even if individual galaxies are mainly held together by 'brown dwarfs' or black holes.

OR ARE WE BEING FOOLED?

The basic evidence for dark matter is that the gas and stars around galaxies move surprisingly fast. This outlying material would be unbound, and would fly out from the galaxy, if it were feeling only the gravity of the galaxy we see. In drawing such inferences, we use our standard theory of gravity, which, in this context, reduces to Newton's inverse square law. This law has been directly tested only within our Solar System; it is plainly a leap of faith to apply it on scales a hundred million times larger. (There have, incidentally, been recent attempts to check the inverse square law in the laboratory, motivated by the idea that some extra 'fifth force' may come into play on scales of a few metres. Here again, the straight experimental evidence is meagre, because gravity is so feeble between laboratory-sized objects.)

Could a different 'long-range' law of gravity obviate the need for extra 'dark' matter? An Israeli physicist, Mordehai Milgrom, conjectures that Newton's inverse square law – according to which the force goes as (mass)/r^2 – goes wrong (and underestimates the true strength of gravity) when the force is weaker than some particular value. This proposal, known as the MOND theory (an acronym for MOdified Newtonian Dynamics), doesn't violate any known experiments or observations, but allows Milgrom to reinterpret much of the data without having to invoke dark matter.

Milgrom has performed a useful service by seriously analysing what it would take to evade the need for dark matter. He has also suggested some tests: for instance, the MOND theory could give rise to patterns of stellar motions in a galaxy that could never arise in a conventional picture because they would require, in some regions, a *negative* density of 'dark matter'. Other tests involve the effect of gravity on light rays. 'Conventional' theories predict how light rays are deflected by any massive object; both 'dark' and luminous matter contribute to this bending. MOND is less specific about this, because the modified gravity may not enhance the bending of light by the same factor as it enhances the force on stars and gas.

Why go to such lengths to avoid having to postulate dark matter? Why should all (or even most) of the gravitating stuff in the universe be shining? There are many forms that dark matter could take, and none seems inherently far-fetched. The challenge is surely to discriminate among the many options, and narrow down the list of candidates. If, at some future date, quests for dark matter had all drawn a blank, and had eliminated all credible options, there might be a stronger motivation for MOND.

Milgrom's proposal seems unappealing for a second reason (in the jargon of consumer magazines, it is my 'worst buy'). It jettisons one of the triumphant successes of physics – Einstein's theory of gravity, which incorporates and extends that of Newton, and has survived amazingly precise tests. The MOND idea, as Milgrom realises, would destroy the entire integrity of Einstein's theory – it is not mere tinkering (or patching up) and would set us back to a pre-Newtonian stage. That would be a high price to pay.

SHORTENING THE CANDIDATES' LIST

It would be specially interesting if some as-yet-unknown kind of particles, left over from the early universe, accounted for the dark matter. But we should then have to view the galaxies, the stars and ourselves in a downgraded perspective. Copernicus dethroned the Earth from any central position. Early this century, Shapley and Hubble demoted us from any privileged location in space. But now even *particle chauvinism* might have to be abandoned: the protons, neutrons and electrons of which we and the entire astronomical world are made could be a kind of afterthought in a cosmos where photinos or neutrinos control the overall dynamics. Great galaxies could be just a 'puddle' of atoms, held in place by ten times as much gravitating stuff in some unknown, and quite different, form.

We don't yet know what types of particle might have existed in the ultra-early phases of the universe, and how many would survive. The answer depends on the laws prevailing at high energies where the physics is still uncertain. When these laws have been clarified, we should be able to predict what 'fossil' particles survive from the first millisecond just as confidently as we can now predict the amount of helium surviving from the first three minutes (Chapter 3). The more dark matter there is, the more the overall cosmic expansion is decelerating: if there is enough, the expansion might eventually halt. Dark matter not only moulds the present structures in our universe, but determines their eventual fate.

Dark matter dominates galaxies. How galaxies form, what they look like and the way they cluster depend on how the dark matter behaved as our universe expanded. We can make different guesses about the dark matter, calculate the outcome of each, and see which most resembles what we actually observe. Such calculations, further discussed in the next chapter, can offer indirect clues to what the dark matter is.

Ordinary atoms may comprise less than 10 per cent of the universe, in terms of mass: cosmic dynamics would be only slightly changed if they weren't there at all. But atoms are plainly a prerequisite for our existence. Without them, a universe could harbour no stars, no chemistry and no (or very little) complexity of any kind. Atoms may be a kind of afterthought, but a universe without them would be sterile.

NOTES

1 Astronomers have developed an elaborate taxonomy for galaxies on the basis of their sizes, shapes and the dominant types of stars in them. Gerard de Vaucouleurs – the world authority on galactic morphology – formulated a scheme with over a hundred categories; and even this failed to include all 'peculiar' systems.

2 Not even the best and sharpest picture of galaxies can pick out individual stars, except for especially bright stars in the closest galaxies to us.

3 X-ray astronomy offers another diagnostic for dark matter in clusters of galaxies, just as it does in individual elliptical galaxies. These systems are pervaded by dilute hot gas, which is manifestly being confined in the cluster by gravity. From its X-ray emission, we can infer the temperature and pressure of the gas. The gas is so hot that it wouldn't be confined at all unless it were being held down by a stronger gravitational force than the stars alone provide.

4 Because the Solar System is orbiting around the Galaxy, and sweeping through the Halo (which itself rotates more slowly than the disc of our Galaxy), impacts would come preferentially from the direction towards which we are moving. This trend could distinguish genuine events from those due to (for instance) radioactivity in the rock. Furthermore, the Earth's speed relative to the average particle in the Halo is not the same

throughout the year, because of our motion around the Sun. The detection rate should therefore vary over the year. Such an annual modulation, with an amplitude of a few per cent and a peak in June, would show that the events were genuine even if we knew nothing about the direction they came from.

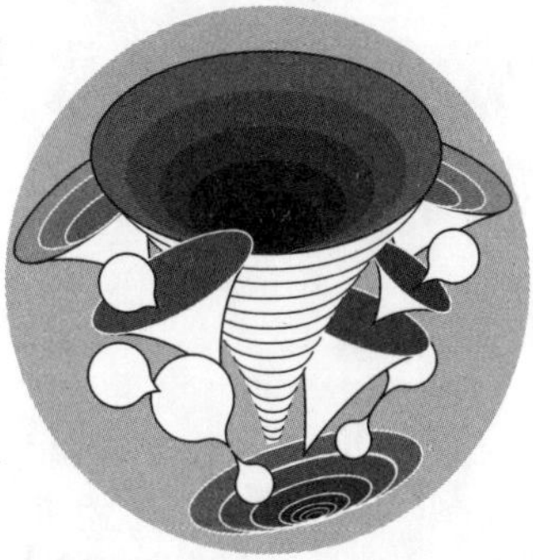

7

FROM PRIMORDIAL RIPPLES TO COSMIC STRUCTURES

An escape from the tight little cage of our universe; tight, in spite of all the astronomists' vast and unthinkable stretches of space; tight, because it is only a continuous extension, a dreary on and on, without any meaning.

D.H. Lawrence

THE HARDWARE

Astronomy was the first 'professional' science (apart perhaps from medicine), even though there has always been a strong amateur interest. It was certainly the first science to use big and expensive equipment. The 18th-century telescopes used by William Herschel

– massive and elaborate constructions – contrast with the modest equipment that would have been used by his contemporaries Levoisier and Cavendish, either of whose 'laboratories' would almost fit on a kitchen table. Astronomy was certainly the first 'big science'. Back in the 16th century, Tycho Brahe's project to map the stars was bankrolled so lavishly by the Danish monarch that he could build a cathedral-like observatory (of which, regrettably, nothing survives) on the island of Hven. The 18th-century expeditions to the Pacific to observe transits of Venus (and thereby fix the size of the Solar System) were costly undertakings by the standards of the time.

With Newton's laws as the sole theoretical input, 18th-century astronomers could compile accurate almanacs predicting the tracks of planets across the sky. The starry firmament was then regarded as a fixed backdrop to the Solar System. The idea that the stars themselves were moving dates from the late 18th century: it was then that the small *relative* motions of stars were first detected. Some stars were realised to be orbiting around binary companions, confirming that Newton's laws held in the celestial realm. But the Sun and stars were still topics for fanciful conjecture. (William Herschel, for instance, despite his sophisticated studies of how stars moved and how they were spread through space, thought the Sun might be inhabited.)

Astronomy received a boost when the photographic plate was introduced in the 19th century: faint objects, undetectable when viewed directly through a telescope, showed up clearly on long-exposure photographs. The bright and beautiful images familiar from books and posters give a misleading impression. The light from galaxies, mainly the smoothed-out contribution from vast numbers of stars, is barely detectable above the glow of the night sky, and only long exposures reveal them at all clearly. When light is split into its constituent colours by a spectrometer it reveals what celestial objects are made of – this realisation, again dating back to the 19th century, began the science of astrophysics.

Throughout the 20th century, techniques improved and telescopes were built with progressively larger collecting areas. By the 1980s more than a dozen had mirror diameters exceeding four metres. Their instrumentation has been upgraded, and made more sensitive and efficient, as technology has advanced.[1]

The only way to detect still fainter objects is by using larger mirrors to collect more light. The first of these 'new-generation' instruments is the Keck Telescope, on Mauna Kea in Hawaii, completed in 1994. This has a 10-metre mirror (actually a mosaic made up of 36 hexagonal elements), and gathers four times more light than earlier telescopes; faint objects therefore show up more clearly.[2] Several other telescopes with 8–10-metre mirrors are being built; and there is a second Keck alongside the first one. But 'Keck one' had a several-year lead over any others.

The Hubble Space Telescope, orbiting high above the blurring and distorting effects of the atmosphere, is still, in some respects, a unique instrument – especially because of the sharp images it provides. But it would have made still more impact had it been launched on schedule in the early 1980s, before there had been such great advances in ground-based telescopes. Because of delays to NASA's space shuttle programme (partly caused by the *Challenger* accident in 1986), its light detectors were 10 years out of date by the time they flew. And, after launch, its images turned out to be badly out of focus, because the main mirror had been incorrectly set up. Experts in 'management' could draw lessons from all that went wrong with this project: effort and responsibility were too diffused; there were too many personnel changes because it lasted so long; no single person had the expertise, authority and commitment to oversee all aspects of the work.

A later shuttle flight, carrying a team of astronauts, visited the Space Telescope when it was already in orbit to replace faulty parts and correct the optics. This 'refurbishment' was presented as a triumph for manned spaceflight: the astronauts performed their complicated mission faultlessly. But, if scientific cost-effectiveness had been the prime criterion, it might have been better to have abandoned the jinxed telescope, and launched an updated copy. Indeed, the entire Space Telescope would have been cheaper, and far less delayed, if it had been decoupled from the manned space programme (and from NASA's space shuttle) right from the start. According to Riccardo Giacconi (who went on from the X-ray astronomy described in Chapter 4 to become the first director of the Space Telescope Science Institute, responsible for operating the telescope in orbit), *seven* similar space telescopes could have been built and launched with separate expendable rockets for what has so far been spent on just one. Even then, each space telescope would still have cost several times more than the largest ground-based telescope.

In parallel with instrumental advances have been those in computing. In less than second, a computer can surpass ten thousand lifetimes of human calculations. These simulations or 'numerical experiments' open up a new dimension. We can compute the detailed consequences of various theories, and see which assumptions yield the closest match with the increasingly precise observations that telescopes now offer us. The disparities between real and imagined universes tell us where our assumptions are wrong, and (with luck) lead towards a more faithful understanding.

THREE PROBLEMS WITH GALAXIES

There are some basic questions about galaxies. One must first ask why such things exist at all. Why are these assemblages of stars and gas the most conspicuous large-scale units in the cosmos? Galaxies have a characteristic size, even though they (like stars) have a broad spread around the average. Is there any physics that singles out galactic dimensions, just as, since the work of Eddington and Chandrasekhar, we have understood the natural scale of stars? To some extent, galaxies must be determined by cosmology – they couldn't exist unless the 'initial' conditions of the expanding universe allowed big enough gas clouds to condense out. However, there is obviously something that determines where in the mass hierarchy *individual galaxies* end and *clusters of galaxies* begin. The Coma Cluster, for instance, consists of about a thousand separate galaxies each containing around 10^{11} stars. But why isn't it, instead, a huge amorphous agglomeration of 10^{14} stars?

Also, there is the embarrassing fact, discussed in the previous chapter, that most of their mass, maybe as much as 90 per cent, is unaccounted for – it's not in the stars and gas that we see, but takes some unknown 'dark' form. Clearly we will never understand galaxies properly until we understand the dominant stuff whose gravity binds them together.

Thirdly there is the fact that some galaxies are more than just stars, gas and dark matter: their main power output comes from a concentration at the centre, probably a massive black hole. These 'active galactic nuclei' raise further problems. Why do some galaxies flare up, and release the colossal amount of radiation that turns them into quasars and radio galaxies? (See Chapter 2 and 5.)

EMBRYONIC GALAXIES

Gravity renders a uniform universe unstable. This is something that Newton himself had realised, at least for a *static* universe. In a letter to Richard Bentley, a classics scholar and contemporary, who was Master of Trinity College, Newton wrote:

> It seems to me, that if the matter of our sun and planets and all the matter of the universe were evenly scattered throughout all the heavens, and every particle had an innate gravity towards all the rest, and ... if the matter were evenly dispersed throughout an infinite space, it could never convene into one mass; but some of it would convene into one mass and some into another, so as to make an infinite number of great masses, scattered at great distances from one another throughout all that infinite space. And thus might the sun and fixed stars be formed ...

In an *expanding* universe, gravity still does essentially the same thing as Newton envisaged. Any region slightly denser than average decelerates more, because of the extra gravity; its expansion lags more and more behind that of an average region, so the density contrast grows. (If we throw two balls upwards with slightly different speeds, their trajectories may, to start with, differ only imperceptably. The slower ball, however, will have completely stopped, and already started to fall, while the faster is still moving upwards.)

A universe that was *completely* smooth and uniform when it started expanding would remain so even after 10 billion years. It would be cold and dull – no galaxies, therefore no stars, no chemical elements, no complexity, certainly no people. But the initial departures from uniformity need only have been very slight: density contrasts are amplified during the expansion, so that even very slight 'ripples' in an almost featureless fireball can evolve into conspicuous structures.

In its compressed early stages, our entire universe was much denser than individual galaxies are today. Galaxies obviously couldn't then have existed as separate entities: their embryos would have been merely regions of slightly enhanced density, whose subsequent expansion was retarded and eventually halted by their excess gravity.

Like the individual galaxies, clusters and superclusters are the outcome of gravitational aggregation. The newly formed galaxies would not have been spread completely uniformly – there would be slightly more in some places than in others. As the expansion continued, any volumes containing an excess mass would suffer extra deceleration, so that the galaxies in those volumes ended up conspicuously more close-packed than average.

There is nothing 'fundamental' about the exact pattern of galaxies in our sky; a decent theory should, however, explain the *statistical* properties of galaxies and their distribution. By analogy, an oceanographer hopes to explain the statistics of ocean waves – their *average* features – but not the detailed pattern of wavecrests shown in a snapshot in a particular place, and at a particular time.

We'd like to interpret our present cosmic environment as a generic outcome of some simple and 'natural' hypothesis about the early universe: Could the galaxies (and the clusters and superclusters) really have evolved from initial fluctuations – fluctuations which (as described in Chapter 10) may have been imprinted when our entire universe was no bigger than a golfball?

UNIVERSES IN A COMPUTER

A region large enough to be a 'fair sample' of our universe contains several thousand galaxies – that amounts to 10^{72} atoms (along, perhaps, with even more particles of whatever kind make up the 'dark' matter). Obviously, no conceivable computer can simulate all the small-scale detail. Fortunately, a 'coarse-grained' simulation suffices if we are interested in the gross properties of galaxies and how they move.

A newspaper picture uses only a few thousand dots to convey a recognisable image of a face. So we may learn something even from a drastically simplified calculation which represents each galaxy by ten thousand 'blobs'. The whole region being simulated might then contain 10^8 blobs altogether. Computers are powerful enough to track at least this many bodies, and to calculate the gravitational force that each feels, due to the pull of all the others.

If blobs started off in an utterly regular pattern, then their evolution would be very simple: all would expand away from each other exactly in accordance with the homogeneous 'model universes' first discussed by Friedmann and Lemaître. But suppose that the blobs were not laid down completely uniformly at the start of the computations. Regions where they were packed just one per cent more densely than average would suffer extra deceleration. By the time the universe had expanded a hundredfold, the excess density would be not one per cent but 100 per cent. These overdense regions would by then have stopped expanding; they would fall together to make galaxies, clusters or superclusters. The universe would go on expanding around them, the underdense regions becoming 'voids'.

The outcome would obviously depend on the details of the initial irregularities – whether there were (say) 10 slightly overdense regions each containing 10 million blobs, or 10 thousand overdense regions each containing just 10 thousand blobs. The first would evolve into 10 big structures, each as big as a cluster of galaxies; the second into 10 thousand small galaxies.[3]

The fluctuations or 'ripples' would have been imprinted very early on, before the universe 'knew' about galaxies and clusters: there would be nothing special about these sizes (or, indeed, about any dimensions that seem distinctive in our present universe). The simplest guess would be that nothing in the early universe favours one scale rather than another, so that the gravitational effect of the fluctuations is the same on every scale. General arguments for such 'scale-free' fluctuations were advanced, in the early 1970s, by the British cosmologist Edward Harrison, and by Yakov Zeldovich. Current theories of the 'inflationary universe' (admittedly still speculative) suggest that the fluctuations would indeed have the 'Harrison–Zeldovich' form, or something close to it. Is such a simple prescription of the early universe consistent with the range of complex structures that have emerged 10 billion years later? This is the question that computer simulations are designed to solve.

A BOX OF 'COLD' OR 'HOT' DARK MATTER?

A typical part of our universe is like an expanding box. At the start, conditions are almost uniform and the box is small and dense; gravity enhances the density contrasts of slight initial irregularities, allowing structures to develop, as the box expands. Can any natural assumptions about the initial fluctuations (and about the dark matter) account for the emergence of galaxies, clusters and superclusters by the time the universe is 10 billion years old? 'Experimental cosmologists' use the fastest computers to evolve a variety of 'model' universes. The results are then displayed as movies: speeded up by 10^{15}, the entire evolution takes a few minutes. The interesting question, of course, is which 'initial conditions' lead to a simulated universe most like our real one.

When data are sparse, it makes sense to test simple and specific assumptions first. It would be astonishing if we hit on the right idea straight away. But observers like to have a specific 'template' to compare their data with (especially if they can refute it); and theorists develop a feel for how the assumptions might be 'tweaked' to fit the observations better.

In this spirit, cosmologists have focused on one particular hypothesis: that the dominant gravitating stuff in our present universe, the 'dark matter', consists of particles left over from the dense early phases. Each of these particles might weigh about as much as an atom, but they would be so weakly interacting – in effect so small – that each one would feel only the collective gravitational effects of all the others, and they would never collide with one another. Well, hardly ever; experimenters are seeking such particles in our own Galaxy by attempting to record the recoil that occurs when, very rarely, one of these particles hits an atomic nucleus – see Chapter 6). The particles would have small random motions, like the atoms of a gas with a low temperature. This hypothesis has hence become known as the 'cold dark matter' (CDM) scenario. It predicts that cosmic structure forms hierarchically: subgalactic scales condense out first; these merge into galactic-mass objects, which then cluster on still larger scales.

An alternative idea is that *neutrinos* make up the dark matter. There is nothing hypothetical about these: we know they exist, and we can calculate how many should survive from the big bang. Neutrions may have small masses, or their masses may be exactly zero – this isn't yet known (see Chapter 6). Neutrinos in the early universe would have high random speeds, just like the atoms in a hot gas – hence the unimaginative term '*hot* dark matter' (HDM). Small-scale irregularities are wiped out (unlike in CDM) because the neutrinos that start in overdense and underdense regions move fast enough to change places, and smooth out any initial fluctuations smaller than superclusters. Very large regions, of supercluster scale, are then the first to form; these afterwards fragment into galactic-mass systems. In other words, structures of HDM form in a 'top-down' fashion, in contrast to the CDM case, where they emerge hierarchically or 'bottom up'.

The simulated universes end up looking very different in the two scenarios. If one resembles our actual universe much more closely than the other, that should be a strong clue to what the dark matter really is. But the comparison actually isn't straightforward. These simulations reliably predict the present-day pattern of clustering *in the 'dark' material* – neutrinos (HDM) in the one case, CDM particles in the other. But we have only indirect clues to how the *dark* matter is actually distributed in our universe. We know far more, of course, about the *bright* matter – made of ordinary atoms – but this could be a misleading 'tracer': galaxies may, for instance, lie preferentially in the largest concentrations of dark matter, just as the white crests highlight the highest ocean waves.

SUPPORT FOR 'COLD DARK MATTER'

The patterns of galaxies in the sky look like a tracery of filaments. But the eye is ultra-efficient at picking out such features, even when they are not significant. (It was better for our remote ancestors sometimes to 'see' tigers that weren't there than to miss the one that really was!) Astronomers react to the patterns rather as to ink-blot psychological tests. They need help from statisticians to describe the clustering quantitatively. But it already looks as though a neutrino-dominated (HDM) simulation yields a smaller range of interesting-looking structures than one with CDM, and bears less resemblance to the actual patterns of galaxies.

A simulation that truly resembles our actual universe must mimic the observed clustering at the present epoch. But it must do more: it must match the data at all the past epochs that we can probe by observing high-redshift objects. Clusters of galaxies are less prominent at high redshifts (that is to say, in the past). This also accords well with the CDM simulation, which show big clusters forming only recently via mergers of smaller ones.

Galaxies (or, at least, their outer halos) would almost touch each other if the universe were five times more compressed. That suggests that protogalaxies stopped expanding and separated from each other at that stage in the expansion, when our universe would have been about a billion years old. *Quasars*, the ultrabright centres of newly forming galaxies, provide important clues to these early stages. When we probe eras back to a billion years after the big bang, we find quasars. Still further away, the quasar population seems to thin out.

Seeking quasars with still larger redshifts is not just a matter of 'breaking the record': success in finding quasars much earlier in cosmic history would push theorists into a corner. Even in the CDM model, about a billion years elapse before the first galaxies formed. The rival 'HDM' theory, which postulates that the dark matter consists of neutrinos with small masses, predicts that galaxies form later than they do in the CDM theory, and already runs into trouble with the quasar observations.

Larger telescopes wouldn't necessarily reveal galaxies further away than those already discovered (even though they'd obviously offer clearer pictures of those that have already been found). If we have already detected the first galaxies that ever formed, looking still further out (and further back into the past) would merely probe a featureless pregalactic dark age.

It is still controversial whether any simple hypothesis about the dark matter (neutrinos? cold particles? or a mixture?) accounts for the entire range of structures in our actual universe.

THE BIGGEST STRUCTURES
A NEW COSMIC NUMBER

Superclusters are only a few times denser than an average volume of space; they are still condensing out from the expanding universe.

On the other hand, smaller-scale systems – individual galaxies and small groups of galaxies – are much more compact, having reached equilibrium a long time ago: there has been time for complicated internal evolution to erase any clear trace of how they formed. The biggest cosmic structures preserve the most direct imprint of their formation, and therefore bear most directly on fundamental questions about the very early universe.

In gravitational terms, even the biggest superclusters are merely slight irregularities on a basically smooth universe. There is a natural way to characterise how strongly a cosmic structure – be it a star, galaxy or cluster of galaxies – is bound together by gravity: we can ask what fraction of its total 'rest-mass energy' (mc^2) would be required to pull it apart. For clusters and superclusters, the answer is about one part in a hundred thousand or 10^{-5}.

Because this important number is so small, gravity is actually quite weak in galaxies and clusters. Newtonian theory is fully adequate for analysing how these structures form and how their internal motions evolve; this greatly simplifies the computer simulations. The smallness of this number also means that we can validly treat our observable universe as approximately homogeneous, just as we'd regard a globe as smooth and round if the height of the waves or ripples on its surface were only $1/100\,000$ of its radius. This number doesn't change with time – it's the same for a slightly overdense region in the early universe as it is for the cluster (or supercluster) that this region eventually evolves into, and characterises the 'roughness' of the early universe. I will call this number Q.[4]

RIPPLES IN THE BACKGROUND RADIATION

We observe galaxies and quasars so far away that their light set out when the universe was a tenth its present age. Conventional astronomy can never, however, probe further back than the era when the first such objects formed and 'lit up' the universe. But, if we're right that cosmic structures emerged via gravitational instability, then their precursors must have existed beforehand as regions with slightly above-average density, expanding slightly more slowly than average. These precursors would leave their trace in the microwave background radiation, itself a relic of the early universe.

Our universe would have started off dense and opaque, like the glowing gas inside a star. Radiation quanta (photons) would scatter repeatedly off the electrons. But after half a million years of expansion the temperature would have dropped to 3000 degrees – somewhat cooler than the Sun's surface. The electrons would then have been moving slowly enough to be captured by protons, forming neutral hydrogen. They then no longer scatter the photons, so the universe would have become transparent. The primeval fog lifts, and the photons can thereafter travel uninterrupted until the present time.

The microwave background radiation reaching our radio telescopes comes, in effect, from a 'last scattering surface' far beyond the most distant quasars – it carries information about an era long before any quasars or galaxies had formed.

This surface – sometimes called the 'cosmic photosphere', by analogy with the Sun's surface, which is called its photosphere – lies, in effect, at a redshift of 1000, this being the factor by which the universe expanded, and the radiation wavelengths stretched, in cooling from 3000 degrees to its present temperature of just under three degrees.

Radiation from an incipient cluster straddling the 'last scattering surface' would reach us slightly cooler than from other parts of the surface, because it would lose energy (and suffer a slight extra redshift) escaping from the extra gravity of an overdense region. The temperature reduction should only be about one part in 100 000 – the small number Q mentioned earlier, which measures how irregular the universe is. This is a challenging target for experimenters on the Earth to aim at: the *total* cosmic background radiation, just below three degrees, is only about one per cent of the emission from the Earth (whose surface temperature is about 300 degrees above absolute zero), and the effect being sought is 100 000 times smaller still.

By 1980 experiments on the ground had become sensitive enough to detect temperature differences across the sky as small as a part in 10 000. But they found nothing but smoothness. A Soviet experiment called RELICT, launched in a satellite to avoid the effects of the atmosphere, scanned the entire sky with somewhat better sensitivity, but still found no non-uniformities. The Soviets were unlucky, because it turned out that only a modest further improvement converted these 'upper limits' into positive results.

COBE AND AFTER

NASA's Cosmic Background Explorer (COBE) satellite was designed to measure temperature differences smaller than one part in 100 000, and succeeded in doing so. To detect this faint signature of the irregular gravity field in the early universe was a technical triumph: it ranks as a major discovery even though the fluctuations weren't unexpected. It would have been more surprising (indeed, we should all have been flummoxed) if fluctuations *hadn't* been detected at COBE's level of sensitivity. An even smoother early universe would have seemed incompatible with the conspicuous clusters and superclusters we see around us today: density contrasts would have needed to grow faster than they do under the action of gravity, and theorists would have been forced to invoke some extra non-gravitational mechanism.

The day after this discovery was announced, in April 1992, I was surprised to find the entire front page of my British daily newspaper filled, under a banner headline HOW THE UNIVERSE BEGAN, with a detailed description of what it meant. The experimenters had convened a press conference, and put out a press release in which NASA-sponsored scientists heralded the measurements in extravagant terms: 'the Holy Grail' . . . 'like seeing the face of God' and so forth. Even Stephen Hawking (who does not depend on a NASA grant) deemed the results 'the greatest discovery of the century, if not of all time'.

Once media attention reaches a threshold level, it feeds on itself and amplifies. It is nothing new for discoveries to be exaggerated and distorted in newspaper headlines: Einstein was himself a victim. But in the COBE case *the researchers themselves* initiated the hype; the media simply reported at face value what 'experts' said. Unfortunately, journalists sometimes need to assess scientists' claims with as much scepticism as they customarily bring to those of politicians.

Science seldom makes 'news'. Scientists can't reasonably complain about this, any more than novelists or composers would complain that their latest works don't make the news bulletins. Important new ideas and discoveries often emerge gradually, through the collective efforts of many people. Journalistic coverage of science that is restricted to 'newsworthy' items – newly announced results with a crisp message that can be easily summarised – can't avoid conveying a distorted impression. That would be true even if the items were optimally chosen: in fact the distortion is even greater because some scientists (and some institutions) are far more effective than others in communicating and promoting their researches.

PUTTING THE PICTURE TOGETHER

The COBE satellite continued gathering data for four years, and mapped the background temperature over the whole sky. Its 'beam' covered an angle of seven degrees, so it blurred over the fine detail. But it probed irregularities on all angular scales from seven degrees up to 90 degrees. The temperature fluctuations were about the same throughout this range – the universe got neither 'rougher' nor 'smoother' as the scale got larger. Stephen Hawking's enthusiasm was so extravagant because some highly promising ideas (the 'inflationary' universe introduced in Chapter 10) predict that fluctuations very like that could have been imprinted when our universe was less than 10^{-36} seconds old. He believed that COBE was telling us about the actual 'quantum birth' of our universe.

The biggest superclusters of galaxies are a few hundred million light-years across – vast dimensions, but still 100 times smaller than our observable universe. If the precursor of such a structure straddled the last scattering surface, it would subtend an angle of only about one degree. COBE therefore probed even larger scales than superclusters of galaxies. Overdense regions on these still grander scales have not yet condensed out to a discernible extent because their extra gravitational energy (only 10^{-5} of their rest-mass energy) cannot compete with the kinetic energy of the expansion, which is more important for larger-scale systems.

Instruments on mountaintops, in balloons or at the South Pole have now detected temperature fluctuations on angular scales of one degree or less. These experiments cannot map the entire sky, as a satellite can, but they achieve the same sensitivity at enormously less expense. Early in the new millenium, however, two new space experiments – one funded by NASA, the other by the European Space Agency – will map the fluctuations in exquisite detail over the entire sky; these should list which (if any) of our current ideas on galaxy formation is correct.

The early universe was smooth in the sense that the ocean is smooth. There's a well-defined average curvature, but there are waves and ripples on it. Looking down from the air on an ocean, you may first see just smoothness, except for the gentle overall curvature of the Earth. But, as your vision sharpens, you begin to discern some waves. A further modest improvement allows you to study the waves in detail. What are the statistics of wave heights? Are the longer waves higher than the shorter ones? This is a metaphor for the exciting potential of microwave background studies. COBE got the first positive result; these have been corroborated and extended; new instruments ten times more sensitive than the present ones are now being built.

THE FABRIC OF OTHER UNIVERSES

The form of the emergent structure depends on a new fundamental constant: the number that measures the 'roughness' of the early universe, the 'height' of the ripples superimposed on the overall smoothness. In our universe this number, Q, is about 10^{-5}. Q tells us the energy in the fluctuations, as a fraction of our universe's total energy. It determines when structures condense out, and the scale of the largest superclusters. Because Q is small, the background radiation is very uniform over the sky, and the simple mathematical idealisation that our universe is, overall, quite uniform is a good enough approximation to be useful.

What would happen in a universe where Q was very different, but where all the physics was otherwise the same as in ours? This is another question that computers can help us to answer. In a much smoother universe (where Q was far smaller than 10^{-5}) no galaxies would ever condense; gas clouds could never cool and fragment into stars. Such a universe would remain dark, pervaded by ever more diffuse hydrogen and helium, even if it continued expanding for vastly longer than 10 billion years.

It is harder to calculate the course of cosmic evolution in a universe that started off less smoothly, with Q bigger than 10^{-5}. Cosmology would certainly be a more confusing subject: a hierarchy of conspicuous structures would extend up to much larger scales than the superclusters in our actual universe; a 'fair sample' would require surveying and averaging over a correspondingly larger volume. But whether any observers (or, indeed, any stars) could evolve in this universe is far from obvious. Violent shock waves would fill the entire space with intense X-rays and gamma rays. Huge gas clouds would condense, much earlier than protogalaxies did in our actual universe, and be heavier and denser – so much denser that they might quickly collapse into vast black holes rather than turning into the kind of stellar aggregate we call a galaxy. Any stars that managed to form would be so densely-packed and fast-moving that close encounters would be frequent. Even if planets could form around such stars, the planetary orbits wouldn't remain stable enough to allow evolution.

PROGRESS AND PROSPECTS

Cosmology used to be derided as a science in which facts were so scarce that theory was unconstrained. That is certainly the case no longer: indeed, rather than there being a lack of facts, it is already a challenge to reconcile all the data with any single scheme.[5]

The microwave background radiation directly probes the era before galaxies formed. It complements what astronomers learn by ever more detailed 'mapping' of our present cosmic environment, and by discovering galaxies and quasars so far away that their light set out when they had only just formed. The challenge is to make sense of this web of evidence, and infer the course of cosmic evolution.

When I'm thinking about how galaxies formed, or about how to explain the puzzles of quasars, my mode of thought doesn't differ in any essential way from that of an engineer trying to devise a 'model' to meet given specifications, or a lawyer assessing forensic evidence. What is distinctive about cosmological phenomena is that they are so remote in time and space, and so different in scale, that we can't confidently extrapolate commonsense intuitions.

Just as lawyers believe there is some truth to be sought, so do cosmologists. This distinction is blurred in the writings of some sociologists, who treat scientific 'belief systems' as social constructs that may be no more than transient myths. An advancing consensus should not be valued for its own sake – only insofar as it signifies genuine growth in understanding.

The sociology and 'politics' of science is in itself fascinating. The style of our research and the relative emphasis placed on different topics is moulded not just by individual personalities, but by the social and cultural climate, and by broader political factors. The willingness (until recently) with which governments supplied huge sums for particle accelerators owed a great deal to the clout that physicists acquired through their key role in the Second World War. Cosmological research has been much influenced (indeed sometimes distorted) by the impetus given to the space programme, which in turn was affected by superpower rivalry; also by the incursions of scientists with different expertise and style. Most of all, our work has been influenced by the opportunities and constraints of the available techniques – experimental, observational and computational. These influences are a fascinating study in their own right. However such studies should not obscure what to us 'in the zoo' seems the essence of science – that it is a collective and cumulative enterprise which, albeit fitfully, is bringing the workings of nature into a sharper and 'truer' focus.

Steven Weinberg has given a nice analogy for the 'realist' view of the *content* of science. 'A party of mountain climbers may argue over the best path to the peak, and these arguments may be conditioned by the history and social structure of the expedition, but in the end either they find a good path to the peak or they do not, and when they get there they know it.'

Cosmologists may still be groping in the foothills. But they have made a remarkable advance. The embryonic precursors of galaxies and larger cosmic structure are no longer just hypothetical entities – we can actually observe them. The fabric of cosmic structure depends on just a few numbers. Any choice of these numbers defines a universe: we feed these numbers into our computer as 'initial conditions' and calculate what changes are brought about by gravity (and other forces) during the expansion. Only a small fraction of these possible universes would lead to stars and galaxies (the complexities within each galaxy then surpass the powers of any computer). Of that fraction, we seek the one that best matches, in a statistical sense, our actual universe. As computer power advances, the resemblances should get more faithful and realistic.

NOTES

1 One of the biggest enhancements came about when photographic plates were replaced by new kinds of light sensors (known as charge-coupled devices, or CCDs). A photographic plate has a 'quantum efficiency' of little more than one per cent – in other words, only about one light quantum (or 'photon') in a hundred actually registers on the photographic emulsion; for CCDs, in contrast, the efficiency is up to 80 per cent. This is not the only reason why ground-based telescopes are now vastly more efficient. Spectra need no longer be taken one at a time: optical fibres now enable several hundred objects within the field of view to be studied simultaneously. Now that only 20 per cent of the photons hitting the mirror are wasted, there is little further scope for improving the efficiency of existing telescopes: hence the resurgence in telescope-building during the 1990s.

2 A second advantage of bigger mirrors is that they allow sharper images. However, even if the mirror quality were perfect, turbulence in the air would blur the images, especially in blue light. In the longer term, astronomers hope to monitor the atmospheric 'twinkling' and instantly correct for it. The US military explored these techniques as part of the 'star wars' (SDI) programme; this work was afterwards declassified.

3 In our actual universe, other forces come into play as well as gravity. A galaxy forms from the collapse of a region hundreds of thousands of light-years across. Its constituent gas radiates away the heat released by its contraction; the gas settles into a disc and fragments into millions of cloudlets each large enough to become the birthplace of stars, thereby initiating the recycling process that synthesises and disperses all the elements of the periodic table.

4 The roughness of the early universe (the 'height' of the ripples, as it were) could in principle depend on scale. However, for Harrison–Zeldovich fluctuations Q is the same for all scales, and has a more fundamental significance.

5 The value of Q inferred from COBE agrees with what is estimated from superclusters, Great Walls and so on. Extrapolations down to clusters, and to the scales of individual galaxies, depend on what the dark matter is. The smaller-scale data are internally consistent with the CDM theory, but only if Q is about half the value implied by superclusters and by COBE. There are several ways of 'tweaking' the standard CDM theory to resolve this discrepancy between smaller and larger scales. The initial fluctuations might, for instance, increase with scale, rather than Q being a universal number. The fit would also be improved if neutrinos have just enough mass to contribute 20 per cent of the dark matter (the rest being 'cold'). Because they would have been fast-moving in the early universe, neutrinos would have tended to smear out and suppress small-scale fluctuations, but to

enhance fluctuations on scales too large for the smearing to be effective. If there were independent experimental support for a neutrino mass in the appropriate range, many cosmologists would happily espouse this 'hybrid' hypothesis for the dark matter. The fit is also improved if there is less dark matter than would be needed to provide the 'critical' density – the density (see Chapter 8) that is required if the universal expansion is to come eventually to a halt.

6 This quotation is from page 149 of Weinberg's book *Dreams of a Final Theory*, (Hutchinson, 1993).

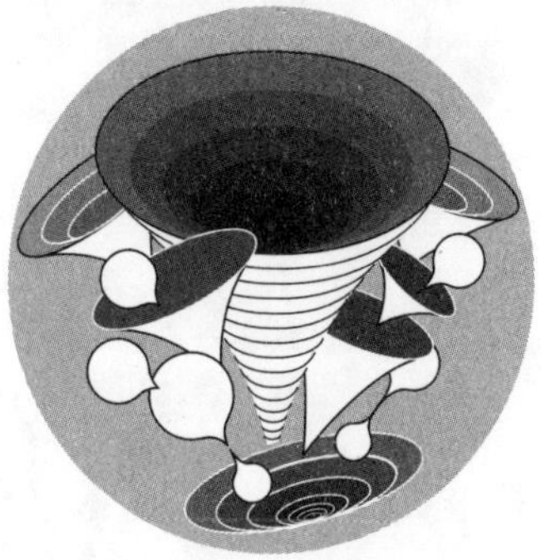

8

OMEGA AND LAMBDA

> The fox knows many things, but the hedgehog knows one big thing.
>
> Archilochus

WHAT IS THE AVERAGE DENSITY?

We've inferred something about our universe's hot dense beginnings, and about how stars and galaxies emerged. But what about the far future? Will everything recollapse so that our descendants all share the fate of an experimenter who ventures into a black hole? Or will the expansion continue for ever?

Our universe would eventually stop expanding, and go into reverse, if gravity were strong enough. Imagine that a big sphere or asteroid were shattered by an explosion, the debris flying off in all directions. Each fragment would feel the gravitational pull of all the others and this would decelerate the expansion. If the initial explosion had been sufficiently violent, then the debris would fly apart for ever; but if the fragments were not moving quite so fast, gravity might bind them together strongly enough to bring the expansion to a halt. The material would then recollapse.[1]

Galaxies are like 'fragments' of our expanding universe. We know how fast they are flying away from us, basically from Hubble's law: what is less certain is the total mass (or the average density) of all the stuff whose gravitational pull brakes the expansion.

It is easy to calculate that our universe would eventually be brought to a halt, and recollapse, if the cosmic density exceeded about *five* atoms per cubic metre; otherwise the expansion should continue for ever. Five atoms per cubic metre is an amazingly low density – far closer to a perfect vacuum than we could ever make on Earth, and a million times more rarified than the gas between the stars in our Milky Way. But our actual universe may be emptier still: if all the stars and gas in all the galaxies were dismantled, and their constituent atoms spread smoothly through space, they would provide only about *a tenth of an atom* per cubic metre – an amazingly complete vacuum, equivalent to a single snowflake spread through the entire volume of the Earth. This is about a fiftieth of what would be needed to bring the universal expansion to a halt. The ratio between the actual and the critical density is normally denoted by the Greek letter omega.[2]

So what is omega? Can it be as large as one? We still can't answer this question. However, we do know that it isn't actually as low as a fiftieth: we saw in Chapter 6 that dark matter outweighs luminous matter tenfold. Whether it is faint stars, black holes or exotic particles, its gravity dominates the 'visible' material in galaxies. However, this reliably inferred dark matter still only brings the density up to around a fifth of the critical density: it contributes an 'omega' of 0.2.

The dark matter that we're confident about is either in the outer reaches of individual galaxies, or in clusters of galaxies (probably having been dislodged from individual galaxies during the relatively close 'fly-bys' that occur within a cluster). Could there be still more spread through huge superclusters, or even pervading the whole of intergalactic space? If there is, it would not affect the motions within galaxies, nor even the motions of galaxies within clusters. But it would slow down the overall cosmic expansion – enough, perhaps, eventually to halt it completely.

Cosmologists betray a strong prejudice, for reasons explained later, that our universe contains the full 'critical density'; this gives an extra motive for seeking the requisite extra dark matter to bridge the gap between an omega of 0.2 and an omega of one.

GREAT ATTRACTORS

Very distant clusters and superclusters are spread more or less uniformly over the sky: insofar as there are equally many in every direction, their gravitational forces on our Galaxy should cancel out. But the 'topography' within a few hundred million light-years, now mapped out in detail, is less uniform. The Virgo supercluster and the Hydra-Centaurus supercluster are not balanced by equally prominent structures in the opposite direction: our galaxy should therefore be feeling a gravitational pull towards them. The strength of this pull would depend on how much dark matter is associated with these superclusters – something that relates directly to omega.

How can we measure our Galaxy's motion? In practice, the best cosmic frame of reference is set by the microwave background radiation, whose effective source (the 'last scattering surface' described in Chapter 7) lies beyond the remotest galaxies, and which therefore averages over absolutely the largest volume we can observe. If we move through this radiation, it will seem hotter in front and cooler behind. Only if we were exactly at rest relative to the average of all very distant matter would the temperature be exactly the same all around us.[3] (This 'preferred frame of reference' is entirely compatible with Einstein's relativity: that theory says the local physics within all freely moving spacecraft is the same, not that the view out of the window is the same from all of them.)

'*Epur si muove*' – 'it still moves, all the same' – were Galileo's words when he reiterated, despite Vatican pressure to recant, that the Earth moved round the Sun. We've known since the 1920s that the Sun is itself orbiting the centre of our Galaxy, but is the entire Galaxy moving? More than 20 years ago, George Smoot and his colleagues looked for the signature of such motion in the background radiation.

They then had no access to space (the COBE satellite came much later) and flew their instruments at high altitude in a U2 aircraft. They discovered that our Galaxy was moving at 600 kilometres per second. We are being pulled towards the Virgo Cluster. But the whole Virgo Cluster is itself being pulled towards more distant clusters, and (in effect) 'pushed' away from voids. Even as far away as 300 million light-years (five times as far as the Virgo Cluster) galaxies are not uniformly spread over the sky, and we are being pulled towards the direction where the concentration is highest.

Our own motion relative to the average 'reference frame' shows up in the microwave background. But detecting analogous motions in other galaxies is more problematic. Alan Dressler's book *Voyage to the Great Attractor* describes how he and six colleagues (who became known among other astronomers as the 'Seven Samurai') discovered that several hundred galaxies, including our own, are being pulled in a coordinated way by some huge mass concentration. Exactly where (or what) this excess mass is remains controversial; but the name Dressler gave it, the 'Great Attractor', has stuck. It is one of the many instances of a catchy name being invented even before the thing it denotes is properly understood.[4]

Speeds of several hundred kilometres per second seem to be quite widespread: our own Galaxy's motion is neither especially high nor especially low. Moreover, the motions are coordinated over surprisingly large scales (rather like a cosmic analogue of plate tectonics here on Earth), so they must be induced by unseen mass concentrations on a very large scale.

The claimed large-scale motions require omega higher than 0.2; they even seem consistent with an omega of 1. There could therefore be even more dark matter, in superclusters and 'Great Attractors' spread through our universe.

COSMIC DECELERATION

The more dark matter there is, the more the cosmic expansion should be slowing down. This *deceleration* should show up when astronomers measure the redshift (or recession speed) of very distant galaxies. Such measurements tell us how fast those galaxies were receding *in the remote past*, when the light now reaching us set out.

Hubble's simple law that redshift and distance are proportional to each other can't be extended unambiguously to large distances (and large redshifts). When the redshifts of very distant objects are plotted against their brightness, the shape of the graph depends on how much deceleration (or acceleration) there has been. Allan Sandage at the Mount Palomar Observatory was the astronomer who did most to push forward the programmes that Hubble began. Using the 200-inch telescope, he was the first person who tried seriously to measure the cosmic deceleration.

Sandage's method, straightforward in principle, was bedevilled by frustrating uncertainties; despite literally decades of effort, it still hasn't yielded a reliable answer. The expansion rate is, of course, changing only very slowly: to discern any changes, one must observe galaxies whose light set out several billion years ago; such galaxies appear very dim, even viewed by the largest telescopes. The galaxies must, moreover, be distinctive and uniform enough to serve as 'standard candles', so that their distance can be inferred from how bright they appear; unfortunately no recognisable class is 'standardised' to better than 20 per cent accuracy.

But the worst problem is that galaxies evolve as they get older. The galaxies crucial for cosmological tests are being seen at less than half their present age: their light has taken 5–10 billion years to reach us. Even if galaxies of a specific recognisable type constituted precise 'standard candles' at the present, one needs to know how each candle changes as it burns.

Any galaxy, in its youth, would have contained many bright stars which by now have died; and all the stars in it would be at earlier stages in their evolution. It seems surprising, in retrospect, that Sandage and others had pressed ahead without properly addressing this issue. The first serious calculations of how galaxies evolve were made in a 1967 PhD thesis by Beatrice Tinsley, a New Zealander working at the University of Texas. From the colours and spectra of well-studied nearby galaxies, she inferred what 'mix' of stars they contained – in particular, the proportions with different masses. Astrophysicists already knew enough about how stars evolve to be able to compute what any particular type of star would have looked like when it was younger. Tinsley added together the evolutionary tracks for stars of different masses, in the proportions appropriate to the nearby galaxies she had observed, and thereby calculated what such galaxies would have looked like billions of years ago. She found that the aggregate stellar population in a galaxy would get dimmer as it aged; young galaxies should be substantially brighter. This wouldn't have been much of a setback if it could be readily calculated and allowed for. But unfortunately the stellar content of galaxies, and the rate at which stars had been forming over their lifetime, weren't then (and, indeed, still aren't) accurately known.

Galaxies 'evolve' for another reason as well. They are not self-contained isolated systems: many seem to be colliding or merging with others.[5] Our own Galaxy will suffer this fate. Andromeda is falling towards it, and in about 5 billion years these large disc galaxies will crash together, the likely remnant being a bloated amorphous 'star pile' resembling an elliptical galaxy. Big galaxies, particularly those at the centres of clusters, are gaining mass, and brightening, by 'cannibalising' their smaller neighbours. This obviously implies that, other things being equal, they would have been fainter in the past.

So a galaxy evolves in two ways: its original stellar population fades as it ages, but it experiences a compensating *brightening* because it gains extra stars from smaller nearby galaxies that fall into it, losing their separate identity. These effects act in opposite directions: either could be large enough to mask the difference between a high and low deceleration, but unfortunately neither can be calculated accurately enough. Direct measurement of cosmic deceleration will remain problematic until we fully understand how galaxies brighten or fade as they age.

Telescopes with modern equipment can reach larger redshifts than Sandage could. But this hasn't helped in pinning down the cosmic deceleration: we then look right back to a time when the galaxies were newly forming, and evolutionary effects are even larger and more uncertain. The Space Telescope, above the blurring of the Earth's atmosphere, offers a sharp enough view of very distant galaxies to show their shape and structure. They are indeed different from nearby galaxies. Many look ragged and irregular; much of their light comes from clouds of glowing gas that have not yet settled into equilibrium or turned into stars.

In the late 1960s, there was a transient hope that quasars (then recently discovered) would greatly aid this task. They are, after all, hyperluminous beacons, up to a hundred times brighter than galaxies, and easily detectable out to very high redshifts. The most distant ones are so far away that their light set out when our universe was a tenth of its present age. Quasars evolve in a very dramatic way. A hypothetical astronomer observing our universe only 2 billion years after the big bang, when galaxies were just forming, would have perceived a vastly more active and dramatic celestial environment. Galaxies were much more prone to indulge in active outbursts when they were young, perhaps because there was then more uncondensed gas available to form and fuel central black holes. But we don't understand this trend well enough to allow for it in the way Beatrice Tinsley could allow for the evolving stellar content of galaxies.

Telescopes are now powerful enough to detect *supernovae* in distant galaxies. The 'stellar bomb' that triggers these explosions has standardised properties – astrophysicists certainly understand it better than the evolution of entire galaxies. These events may provide the 'standard candles' that have hitherto been frustratingly lacking.

IS OUR UNIVERSE OLDER THAN ITS OLDEST STARS?

If the expansion were being strongly decelerated, our universe would have expanded much faster in the past, and the big bang wouldn't have been quite so long ago. Perhaps, therefore, a high omega can be ruled out because a big deceleration would imply a universe younger than the oldest stars in it – an obvious absurdity.

The heaviest and brightest stars burn their hydrogen fuel fastest and have the shortest lifetimes. If a group of stars formed at the same time, those much heavier than the Sun would burn up relatively quickly, those like our Sun would die after 10 billion years, and stars with 0.7 times the Sun's mass would still be shining after 15 billion years. Astronomers can therefore 'date' a group of coeval stars by determining the mass of the heaviest surviving stars – the lower this is, the older the group must be.

Deducing any star's mass from its appearance isn't straightforward. The combined 'fuzziness' of the theory and observations introduces some elasticity into the age estimates. Many experts would nevertheless stake their scientific reputations (and real money as well) on some groups of stars being at least 12 billion years old – pushing the ages still lower strains several cherished beliefs about stellar structure. The time since the big bang must therefore be larger than this.

If the cosmic deceleration could be ignored (in other words, if omega were very small), the time since the big bang would simply be the present distance of a galaxy divided by its present recession speed. (Hubble's law implies that the speed is proportional to distance, so this 'Hubble time' is the same whatever galaxy is used in doing this simple sum.) But to calculate how long a journey takes, you must divide the distance travelled by the *average* speed. If the deceleration were just enough eventually to stop the expansion (a universe with an omega of one) the average expansion speed would be 3/2 higher than it is now: the age of the universe would then be only 2/3 of the Hubble time. The oldest stars therefore tell us something very important about cosmology; perhaps they rule out a high omega.

This line of argument would be very effective if we knew the Hubble time. The recession speeds are known: the redshift is easy to measure. But to infer the Hubble time we need the *distances* of galaxies as well. And these distances are still frustratingly contentious.

COSMIC DISTANCES AND THE HUBBLE TIMESCALE

Cosmic distances are measured by a network of interlocking methods. Some stars (the nearest) shift their positions in the sky when observations are made six months apart: this simple 'parallax' effect, induced by our changing vantage point as the Earth moves around the Sun, tells the distance of nearby stars. This baseline then allows us to determine the distances of more remote objects by comparing the relative brightness of 'standard' objects. For example, if one star's distance is known from parallax methods, then an identical star that appears four times fainter must (from the inverse square law) be twice as far away. We can thus infer the distance of the latter star even though it may be too far away for its parallax to be measured. The distance ladder can be extended outwards, step by step, using a succession of intrinsically brighter objects.

But such procedures introduce cumulative errors and uncertainties. Despite all efforts, measurements of the Hubble time until recently still ranged from 12 to 20 billion years. (It is embarrassing that such a fundamental number is so uncertain, even though other astronomical quantities are known to 15 decimal places.)

The most famous distance calibrators are stars known as Cepheid Variables. These stars are passing through an unstable phase in their life cycle, and pulsate with a period of a few days or weeks. The period depends on the intrinsic brightness in a known way – the brighter stars pulsate slower. The importance of Cepheid Variables stems from their being bright enough to be seen in the closer external galaxies.

One of the Space Telescope's so-called 'key projects' is to use Cepheid Variables to determine the distances of galaxies. Although its mirror is smaller (2.4 metres in diameter) than many telescopes on the ground, it can detect these stars out to greater distances. This is because light rays reaching a ground-based telescope are smeared by the irregularities in the Earth's atmosphere. The Space Telescope concentrates the light from a faint star into an image ten times sharper; very faint stars therefore show up more clearly against the faint diffuse background light from the whole sky. When the Space Telescope has observed Cepheid Variables in enough galaxies (including several as far away as the Virgo Cluster, our nearest big cluster of galaxies) it ought to reduce the present uncertainty in the cosmic distance scale to no more than 10 per cent.

The uncertain steps in the 'distance ladder' could be bypassed if we could actually *calculate* the intrinsic luminosity of some kind of very bright object visible at great distances: there is real hope of understanding supernovae well enough for this purpose.

Yet another approach to the distance scale was first suggested, as early as 1964, by the Norwegian cosmologist Sjur Refsdal. This involves *gravitational lensing*. A galaxy along the line of sight sometimes produces *multiple images* of a very distant quasar: rays are bent by the galaxy's gravitational field, so that light from the quasar reaches us via two (or more) different paths. The path lengths would be unequal unless the alignment between the quasar and the lensing galaxy were absolutely perfect and symmetrical. The paths differ in length by only a tiny fraction, because the light is deflected only very slightly: one light path might, for instance, be longer than the other by one part in 5 billion. Just from knowing the *angles*, we can calculate the *fractional* difference between the path lengths; but we need some other information to set the absolute scale.

This crucial extra information comes from the time-lag between light signals traversing the two paths. Quasars are unsteady in their brightness: they flare and fade irregularly, on timescales of days, weeks or years. Whenever a lensed quasar flares up, one image (the one corresponding to the shortest path) brightens before the other. If, for example, the paths differed by one part in 5 billion, a time lag of one year would imply that the quasar was 5 billion light-years away; a delay of two years would mean the distance was 10 billion light-years. By measuring the time delay between signals reaching us along slightly different paths, we can directly determine the distance of an exceedingly remote object, bypassing all the uncertain intermediate steps in the 'cosmic distance ladder'.

If the Hubble time turned out to be as much as 20 billion years, then even two-thirds of it, corresponding to the age of our universe if omega is 1, might be comfortably longer than the age of the oldest stars. (For real comfort, there should be at least 1 billion years between the figures, since the universe would probably have been at least that old before any galaxies formed.) On the other hand, the lower end of the range – a Hubble time of 12 billion years – would be an embarrassment even if there were no deceleration.

Neither the age of the oldest stars nor the Hubble time is yet pinned down well enough for us to be comfortable with any specific set of numbers. But the uncertainties are now falling to the ten-per-cent level, with as yet no glaring contradiction. And there is real hope that new techniques will narrow the present uncertainty in omega. If I were to place a bet now, it would be that omega will indeed equal 1 (the matter mainly being in some as-yet-unknown particles), and that the Hubble time will be long enough to obviate any serious conflict with the ages of stars. I am also consoled by Francis Crick's adage that no theory should agree with all the data, because some of the data are sure to be wrong!

EVIDENCE FROM HELIUM AND 'HEAVY HYDROGEN'

One way of estimating omega is to compile a complete inventory of dark matter; another is to measure the cosmic deceleration. A separate line of argument puts a ceiling on the amount of gravitating stuff that is made up of *ordinary atoms* – whether in stars, dark objects or diffuse gas – rather than exotic forms of dark matter. This constraint stems from the fact that nuclear reactions in the early universe were sensitive to how close-packed the atoms were at that time.

In the few minutes it took for our universe to cool from 10 billion to 1 billion degrees, nearly 25 per cent of the primordial material was converted into helium (as described in Chapter 3). One of the intermediate products in helium formation is deuterium. This has a nucleus containing a single proton (like hydrogen), plus (unlike hydrogen) a neutron as well – hence the alternative name 'heavy hydrogen'. The more atoms there are in each volume of space, the higher the chance of collisions, and the less likely it is that deuterium nuclei survive the first few minutes without being processed all the way into helium.

Deuterium is observed in interstellar gas, and in the Solar System. Its signature is even seen in the spectra of distant quasars. It cannot be made in stars. Indeed, when a new star condenses out of interstellar gas, any deuterium inside it would be destroyed (converted into helium) before the star gets hot enough to start burning ordinary hydrogen. Deuterium is, like most of the helium, a fossil of the big bang. When the entire universe was at a billion degrees, hot enough for nuclear reactions to occur, the nuclei therefore cannot have been too densely packed – otherwise the deuterium would all have been processed straight into helium and less would survive from the big bang than we actually observe.[6] Although this argument refers to processes in the first few minutes, it involves no exotic physics. Even at that early era, the primordial material was no denser than air, and the individual atomic nuclei are moving at speeds that can be readily reproduced in the laboratory. (It is only in the first millisecond – as discussed further in Chapter 9 – that the primordial material was in an exotic hyperdense state.) The reactions that made deuterium and helium are all well tested experimentally; and the methods whereby astronomers measure the proportions of these elements are very standard and traditional.

The important conclusion[7] is that ordinary atoms cannot contribute more than a tenth of the critical density – in other words they cannot contribute more than 0.1 to omega. This would just, by pushing the numbers, permit the dark matter in individual galaxies to be in some form such as low-mass stars. But the large-scale motions of galaxies in superclusters, and towards the Great Attractor, require omega to be at least 0.2 (even though there is no clear evidence that it is as large as 1).

The more widely dispersed dark matter there is, the less likely it is to be made of ordinary atoms. Otherwise one of the 'pillars' of the big bang – the agreement with the abundances of helium and deuterium – would crumble.

IS THE UNIVERSE FLAT?

In this chapter, and earlier in Chapter 6, I have outlined the case that our universe contains more than we see. Unless we jettison Newton's laws (and I admitted a strong bias against this), we are forced to conclude that galaxies are being pulled by large-scale gravitational forces induced by dark matter, which probably consists of exotic particles rather than ordinary atoms.

We don't know whether there is enough dark matter to contribute the full critical density (omega equal to 1); but there are some general 'philosophical' reasons for favouring this option. If omega were (say) 2, our universe would stop expanding completely after doubling its scale, and would then recollapse. On the other hand, if omega were 0.5, then the expansion would continue, but gravity (and the consequent deceleration) would become less and less competitive with the expansion: when the universe had expanded by a factor of 2, omega would have dropped to 0.25.

At least we know that omega doesn't differ enormously from 1 today – there is not a tremendous imbalance between the effects of gravity and the energy of expansion – and this has striking implications for the early universe. What would omega have been when the helium and deuterium were made – when our universe was a few seconds old? Any deviation from unity would have amplified during the expansion – if omega had started off less than 1, kinetic energy would quickly gain the upper hand, and omega would fall towards zero; if omega were substantially larger than 1 gravity would soon have stopped the expansion completely. The fact that omega still, after 10 billion years, hasn't diverged wildly from 1 means that, when our universe was a second old, omega could not have differed from 1 by more than *a part in* 10^{15}.

Our universe started off with an incredibly fine-tuned balance between expansion and gravity. (The same argument becomes even stronger if we extrapolate back to still earlier stages.) This leads one to suspect that the 'tuning' may be exact – even if we aren't yet sure why. Otherwise it would seem a coincidence that we are observing just at the era when the deviation from flatness first shows up. This recalls an argument from a different branch of physics, concerning the mass of the photon: experiments set a very low bound, less that 10^{-55} gms; but most physicists would bet strongly that this number, known to be so close to zero, is actually *exactly* zero, for some deep theoretical reason.

These prejudices are bolstered by a general suspicion that the entire universe has *exactly zero net energy*. Everything has energy, simply from the famous equation $E = mc^2$. But any bodies influenced by their mutual gravity also have some *negative* energy (called their gravitational 'binding energy'). For example, it takes energy to lift something from the Earth's surface into deep space; conversely, if something falls on to the Earth, this amount of energy is released. What is the gravitational binding energy of a particle due to everything else in our universe? Simple estimates show that it is comparable with mc^2, the rest mass. It is an appealing conjecture that these energies (of opposite signs) are precisely equal: it would then have 'cost nothing' to create all the matter.

Theorists developed a different 'mind set' during the 1980s. This happened partly because the possibility of exotic particles left over from the hot early universe came to be taken much more seriously, and seemed almost a natural expectation. But the main new element was the concept of an inflationary universe, described in Chapter 10, according to which the universe grew exponentially at a very early stage. It would have been 'stretched flat', becoming far larger than the scale of our present horizon, and the kinetic and gravitational energies would be in almost exact balance. This concept is very appealing, and resolves some well-known and stubborn cosmological paradoxes in a natural way: it instils a strong prejudice that, when we add together all possible forms of matter in our universe, we shall reach exactly the critical density.

COSMICAL REPULSION AND 'LAMBDA'

Soon after he developed his theory of gravity, general relativity, Einstein realised that it allowed a *static* universe, now known as the 'Einstein universe'. This universe is finite, eternal and unbounded: a light ray comes back to its point of origin, and starts a new circuit, after a definite time that depends on what the density is. For this universe to obey his equations, Einstein had to introduce an extra complication. This was the so-called 'cosmological constant', denoted by the Greek letter lambda, which gives rise, even in empty space, to a repulsive force proportional to distance. A universe can remain static if the cosmic repulsion exactly balances the gravitational attraction of the matter (which, as in all these idealised cosmologies, is assumed to fill space smoothly and uniformly).

When our actual universe was found to be expanding, Einstein described the introduction of lambda into his equations as 'the biggest blunder of my life'. It had prevented him from *predicting* the expansion that Hubble discovered; Alexander Friedmann had by then already realised that lambda wasn't needed in an expanding or contracting universe.

Einstein's original motivation for introducing lambda was quickly subverted. But that does not discredit the concept that empty space itself may affect cosmic dynamics. Indeed lambda was later resuscitated in a more modern guise, as the 'energy of the vacuum'. According to Einstein's equations, a vacuum energy would cause a 'gravitational repulsion', and an acceleration of the cosmic expansion, exactly equivalent to the old lambda. This is quite opposite to what happens when the same energy is in a more familiar form.[8]

We know that lambda is very small. The cosmic repulsion doesn't discernibly influence orbits in our Solar System, nor even motions on the scale of the galaxies. It can at most be competitive with the gravitational effects of diffuse intergalactic matter – important for the overall cosmic expansion, but not within individual objects. Theorists who worry about the structure of space on very tiny scales are perplexed by why lambda isn't *huge*, similar to the density of the universe at the very early epoch when the particle masses and forces were imprinted. Some process must suppress it, or cancel it out. But does it then become exactly zero? Ideas about why lambda is small are still exceedingly speculative (they invoke 'wormhole' connections to other universes, and similar concepts). Moreover, as we see in Chapter 12, future transitions could make it lower (or even negative) with catastrophic consequences.

As a universe expands, the density of ordinary matter drops off. Unless lambda is exactly zero, it eventually takes over and causes the expansion to accelerate. It could be that individual clusters of galaxies, with their stars, gas and dark matter, are a fair sample of what the universe is made of, and there is not enough dark matter, anywhere, to make omega much more than 0.2. Theorists who required a 'flat' universe where the overall cosmic mass–energy density was exactly 'critical' would then have to introduce the vacuum energy to 'balance the books'. Part of the unease about this is that it seems a coincidence (or yet more fine tuning) that the cosmic repulsion should just be taking over at the present era – the transition being neither in the remote past nor in the far future.

Lambda has interesting (and appealing) implications for the age of our universe. It makes the expansion accelerate – that is to say, it has the opposite effect to ordinary material (stars, gas and dark matter), whose gravity tends gradually to slow the expansion down. Our universe's average expansion speed could then have been *less* than its present speed, and its age *greater* than the Hubble time. In contrast, without lambda the age is always less than the Hubble time, by an amount that depends on omega.

If the Hubble time turns out, when measurements are more precise, to be at the short end of the currently allowed range, then lambda could rescue us from the conundrum that stars seem older than the universe. Most cosmologists would regard this as an 'ugly' solution – they would rather not introduce an extra (non-zero) number into cosmology. But a historical analogy suggests that this prejudice may be short-sighted. Galileo was unhappy when Kepler showed that the planets moved in ellipses, not perfect circles: it seemed less 'natural', certainly less elegant. Newton later showed that ellipses were the natural consequences of his law of universal gravitation. Newton's deeper explanation would surely have reconciled Galileo enthusiastically to Kepler's ellipses. We may, likewise, when we better understand the laws governing the multiverse, realise how narrow and distorted our earlier perspective was. The previously favoured 'flat' universe with zero lambda may then seem special and atypical, just as circular orbits have seemed since Newton.

The multiverse may encompass universes with all possible values of lambda. Galaxies could never form in a universe where lambda was too large: the cosmic repulsion would take over before they had condensed out. So, even if lambda is large in a typical universe, we need not be surprised that we are in one where it is low.

WHAT LIES BEYOND OUR PRESENT HORIZON?

Will our descendants have an infinite future? Or will everything be engulfed in a universal squeeze to a 'big crunch'? These are the contrasting 'futures' that cosmologists have generally discussed (on which it would still be prudent to hedge one's bets). The answer depends on omega, the average density divided by the critical density. Very few cosmologists now favour a value as high as (say) 2, as would be required if the universe were to recollapse within the next 100 billion years. The issue is whether the average density is well below the critical value (so that omega is around 0.2), or is so close to critical that, even if the 'crunch' eventually comes, the build-up to it would be less dramatic because all stars would already have died, and maybe even all atoms would have decayed.

If the big crunch were destined to happen within (say) 100 billion years, we should be in a 'closed' universe whose total content was only a few times larger than we can see. About a fifth of the total volume would already be in view; so only a limited extrapolation is involved in inferring what the rest is like. If we have already seen about halfway around the universe, it is plausible to conjecture that the other half is similar and meshes smoothly on to the part we see.[9] But if enough time eventually elapses for light to arrive from *far* beyond our present horizon, cosmic domains differing drastically from our presently observable universe may come into view.[10]

If our universe goes on expanding for vastly longer, everything we can now observe – what we now call the observable universe – will eventually seem merely a local patch, tiny compared with the vaster horizon to which an observer's view would then extend. Light from galaxies far beyond our present horizon will eventually have time to reach us. This prospect then raises a fundamental question. Is the 10-billion-light-year patch we can now observe typical of all the rest? Would our hypothetical remote descendants just see more of the same?[11] It would be surprising if the cosmic scene changed drastically *just beyond* the limits of our present observation – just as, if you were at a random place in the ocean, it would be surprising if a continental shore lay just half a mile beyond the horizon.

The time already elapsed since the big bang is no more three times longer than the age of the Earth. Although cosmology involves distances vastly exceeding terrestrial ones, the *timespans* of cosmic history are not much larger than those geologists are used to. But perhaps that means that, timewise, we are still near the beginning. If omega is close to the critical value of 1, but exceeds it very slightly, our universe has aeons stretching ahead, even if it is fated eventually to recollapse.

NOTES

1 One might feel somewhat uneasy about applying a result based on Newton's theory of gravity to a whole universe. But even though one cannot describe the global properties of a universe properly, nor the propagation of light, without using a more sophisticated theory such as Einstein's general relativity, the dynamics of the expansion *are* more or less the same as those given by Newton's theory.

2 The critical density of 5 atoms per cubic metre quoted here corresponds to a Hubble time t_H of 15 billion years; the actual critical density depends on $1/t_H^2$. Fortunately, many techniques for measuring cosmic densities scale with t_H in the same way. Uncertainties in the Hubble time, mentioned later in this chapter, do not necessarily introduce corresponding uncertainties into estimates of omega. (Unfortunately, these densities are hard to pin down for quite other reasons!)

3 Only in that frame would we see the very distant universe expanding isotropically away from us. Any other observers would see smaller redshifts in the direction they're moving towards, but bigger redshifts in the opposite direction.

4 What we really need is some technique for revealing concentrations of dark matter on supercluster scales that bypasses all the uncertainties about galactic motions and distances, and about how galaxies are related to dark matter. One new approach involves searches for distortion of the images of very distant galaxies due to gravitational light-bending by superclusters along the line of sight. Such distortions have already been seen in galaxies lying behind clusters (see Chapter 6). The dark matter in superclusters would be less concentrated than in a cluster, and therefore could not

create such strongly distorted or magnified images. But superclusters are such large features on the sky that there may be hundreds of thousands of faint galaxies lying behind each one. These would be distorted in a correlated way, so even a distortion amounting to only a few per cent could be detected statistically if it affected all the background galaxies in the same way.

5 The spacings between galaxies are only ten times their overall sizes – even less in clusters of galaxies. In contrast, the individual stars within each galaxy are spaced by millions of times their actual sizes. Galaxy collisions are not unduly rare; on the other hand, stars rarely crash into one another, except maybe in the central core of galaxies where they are packed especially closely.

6 This constraint on the density of nuclei in the early universe straightforwardly yields a limit on the *present* density of atoms, because the amount of expansion (and consequent dilution) is directly related to the amount the temperature has dropped. The present temperature is just below 3 degrees above absolute zero. In cooling from 3 billion degrees, the universe would have stretched by a billion (10^9), and its density would have dropped by 10^{27}. These estimates can be refined by allowing for the destruction of deuterium in stars, and by fitting the abundance of helium and also of lithium (a rare element which, like deuterium and helium, is believed to be primordial).

7 The precise constraint depends on the Hubble time t_H. The number quoted here is a generous upper limit. There have been attempts to evade this conclusion, or at least weaken it, by supposing that the baryons in the early universe were 'clumpy' rather than uniformly distributed. This is an ad hoc remedy, without any theoretical motivation. It turns out not to widen the allowed range of densities very much anyway.

8 This concept of vacuum energy will reappear in Chapter 10 in connection with the 'inflationary' phase in the ultra-early universe. According to this theory, the vacuum originally had a very high energy (equivalent to a very large lambda). In that perspective, there seems no objection to a non-zero lambda now – indeed the mystery is why it isn't *unacceptably* large.

9 From a mathematical viewpoint, it is even possible that our universe is 'multiply connected'. What this would mean is that our locality could expand for ever, even though we were in a 'small' universe where we see the same finite volume over and over again in a lattice structure, like a kaleidoscope. Until recently, this option, odd though it might seem, couldn't be readily tested However, we can now exclude a 'cell size' smaller than the part of the universe we can presently observe. The evidence comes from noting that we don't see duplicates of conspicuous structures like clusters and great walls. Also, the non-uniformities in the background temperature detected by COBE reveal waves on all scales up to the Hubble radius: there is no evidence that they are truncated at any putative cell size smaller than this.

10 In universes without lambda, the expansion decelerates, because of the gravitational force that everything exerts on everything else. If the density is indeed near the critical value (in other words, if omega is one), the deceleration law turns out to be especially simple: when the galaxies

have moved 4 times further apart, the universe will be 8 times older than it is now (not just 4 times, as would be the case if the expansion didn't slow down at all). Any expanding sphere has a gravitational energy that goes as 1/(radius), and a kinetic energy that depends on the square of its expansion speed. When omega is 1, the net energy of any sphere 'carved out' from the expanding universe is exactly zero, the (positive) kinetic energy being equal to the (negative) gravitational energy. If the universe expands by a factor of 4, the gravitational binding energy of any sphere goes down by 4. The kinetic energy, proportional to the square of velocity, must therefore go down by the same factor, which means that the expansion speed drops by a factor of 2. The age of the universe (radius/velocity) has therefore gone up by 8, and so has the distance that light has been able to travel (that is to say, the distance of the horizon). However, at that stage, the galaxies in the decelerating universe are only 4 times further away. Everything visible to us today will therefore, at that remote future date, be less than halfway to the new horizon. So the new horizon will encompass $2^3 = 8$ times more galaxies.

11 The background temperature is the same, to within 1 part in 100 000, in whatever direction we look. Deviations from homogeneity are therefore small on the scale of our present horizon. We can even infer something about 'lumps' or waves on scales somewhat beyond the Hubble radius, because the resultant gradient would make the background radiation hotter on one side of our universe than the other. But observations set no constraint at all on scales a thousand times larger than our present horizon, because these would create an undetectably gentle gradient across the domain we can observe.

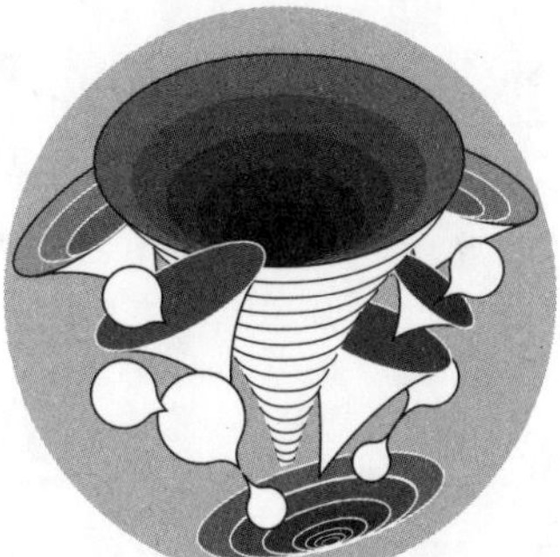

9

BACK TO 'THE BEGINNING'

With increasing distance our knowledge fades and fades rapidly. Eventually we reach the dim boundary, the utmost limits of our telescope. There we measure shadows, and we search among ghostly errors of measurement for landmarks that are scarcely more substantial. The search will continue. Not until the empirical resources are exhausted need we pass on to the dreamy realm of speculation.

Edwin Hubble
Realm of the Nebulae (1936)

STARTING WITH A SIMPLE BIG BANG a fraction of a second old we're learning how, over 10 billion years, our present complex cosmos could have emerged. But why was our universe 'fine-tuned' to expand at a rate that is, seemingly, so precisely adjusted? Another surprise is that its expansion rate (the Hubble constant) is the same in all directions. Why was our universe endowed with the observed mix of atoms and radiation? Why, despite the overall smoothness, does it contain the fluctuations or 'ripples' without which those atoms would have remained as cold featureless gas? Why are the simple 'cosmological models' (dating back to Alexander Friedmann in the 1920s) such an astonishingly good approximation? Why does our universe have the uniformity that is a prerequisite for progress in cosmology?

All these are indeed fundamental questions. But there's nothing new about *that sort* of question. Let's go back 300 years, to Newton. He showed why the planets traced out ellipses as they moved around the Sun – something that Kepler had earlier discovered, but which seemed mysterious until Newton proved it was the direct outcome of his inverse square law of universal gravity. But it remained a mystery to Newton why the planets were 'set up' with their orbits almost in the same plane. In his *Opticks* he writes: 'Blind fate could never make all the planets move one and the same way in orbits concentrick ... Such a wonderful uniformity in the planetary system must be allowed the effect of choice.'

This coplanarity is now understood – it's a natural outcome of the Solar System's origin from a spinning dusty disc that condensed into planets (see Chapter 1). But the demarcation between phenomena that are the outcome of known laws, and those that are mysterious 'initial conditions', still exists as sharply as it did for Newton – we are still, at some stage, reduced to saying, 'Things are as they are because they were as they were.' Progress has pushed the barrier back enormously earlier than the beginning of our Solar System – right back, indeed, to the state when our entire universe was a 10-billion-degree 'fireball' that had been expanding for just one second, and uniform except for slight 'ripples' which condensed, very much later, into cosmic structures.

Cosmologists often get asked how the latest advances relate to philosophy or religion. My personal response is a dull one. Despite all we are learning about our cosmic environment, I don't think the interface with philosophers and theologians is, in principle, any different from what it was in Newton's day.

But can we push the barrier further back than the first second of the big bang, into the first *tiny fraction* of a second?

THREE COSMIC NUMBERS

When the time on the 'cosmic clock' was one second, a 'recipe' for our universe would have required just a few numbers.

- The first specifies the average density in *all* forms of matter, luminous and dark. This number, 'omega', was discussed in Chapter 8; it measures the balance between gravity and kinetic energy, which controls our universe's ultimate fate.
- We also need to know what form this matter takes. The number of protons and neutrons (collectively called 'baryons') is conveniently measured by relating it to the number of *photons* (quanta of radiation), which is accurately known from the COBE measurements of the background radiation. The ratio is very small: there is about one baryon for every billion photons. But a universe without this admixture of baryons – the ordinary atoms that make up the stars and gas in galaxies – would be dark and featureless, quite unlike our own. The baryons contribute to omega, but other kinds of particles left over from the big bang may make a bigger contribution: if so, they are part of the recipe too.
- Finally, we need to specify the amplitude of the primordial 'ripples' that evolve into galaxies, clusters and superclusters. This number, *Q*, is about 10^{-5} (see Chapter 7).

These numbers suffice to determine the main features of our present universe. Once they are specified, we can in principle run a computer simulation of cosmic evolution, ending up when the universe has cooled to 2.7 degrees (see Chapter 7): after 1 billion years, galaxies condense; they later agglomerate into clusters and still larger cosmic structures. We can compare the simulated 'universe in a computer' with what astronomers actually see.

The starting-point of the simulation is a universe about a second old, specified by just three numbers. Can we understand these numbers, in terms of some still more basic principles, by pushing the barrier back earlier still?

BACKWARDS INTO THE FIRST MICROSECOND

The everyday world is governed by atomic structure and gravity; the same well-tried physics (plus that of atomic nuclei) is sufficient for our understanding of ordinary stars – indeed it suffices for extrapolating the big bang back to the time when it was one second old. But when we extrapolate further back to still hotter and denser phases – the first millisecond, the first microsecond and so on – conditions become more extreme, and we can't be confident that the physics we know, and can test in the laboratory, is either adequate or applicable. (Similar uncertainties confront us in, for instance, black holes at the centres of galaxies.) The three 'cosmic numbers' are all legacies of the very early stages. Their values, and perhaps the physical laws themselves, were 'laid down' when conditions were so extreme that the physics is still speculative.

During the first millisecond, everything would have been squeezed denser than an atomic nucleus, or a neutron star. Particles would have large random 'thermal' speeds, because of the high temperatures, and would repeatedly collide with each other. The most powerful particle accelerator, the Large Hadron Collider (LHC) now being built at CERN in Geneva by a consortium of European nations, will be able to simulate the energies that prevailed when our universe was 10^{-14} seconds old – but at still earlier times every particle in our universe carried more energy than even that monstrous machine will generate.

The interval between 10^{-15} seconds and 10^{-14} seconds is likely to have been just as eventful as that between 10^{-14} seconds and 10^{-13} seconds, even though the former is 10 times briefer (and likewise for even earlier intervals). It is therefore more realistic to give equal weight to each power of 10. In this perspective there is plenty of action at even earlier stages – to ignore these early eras is a severe omission indeed.

Theoretical physicists are attempting to interrelate the basic physical forces that govern the cosmos and the microworld. To quote again from Newton's *Opticks*: 'The attractions of gravity, magnetism and electricity reach to very sensible distances, and so have been observed by vulgar eyes, and there may be others which reach to so small distances as hitherto escape detection.' Michael Faraday and James Clerk Maxwell in the 19th century showed that electric and magnetic forces were interrelated – they unified them into 'electromagnetism'. This left just two forces: electromagnetism and gravity. But these are, we now know, supplemented by two further forces which could not be 'observed by vulgar eyes', because they act only at very short range within atomic nuclei: the 'strong' or nuclear force (which binds the protons and neutrons together in atomic nuclei, and overwhelms the disruptive electrical repulsion between the protons), and the 'weak' force (important for radiative decay and neutrinos).

Physicists would like to discover some interrelation between all four of these forces – to interpret them as different manifestations of a single primeval force, just as electric and magnetic forces are interlinked. It was because he took no cognizance of the nuclear and 'weak' forces that Einstein's attempts at unified theories, throughout the last 30 years of his life, were doomed to fail.

The first modern step towards unification is associated with two great names in modern physics, Abdus Salam and Steven Weinberg, who worked independently, each building on the work of other theorists. Salam is a Pakistani (and devout Muslim). His unique legacy (over and above his scientific insights) has been the International Centre for Theoretical Physics in Trieste, which he founded and directed. Its purpose is to remedy the intellectual isolation of scientists in the developing world, and thereby reduce the pressure on them to migrate (as Salam himself did) from their home countries. Such people get the chance to spend part of each year in Trieste, in the stimulating atmosphere of a large international institution.

Weinberg has already been quoted in Chapter 7: he has, in addition to his specialist work, written eloquently on general scientific issues; his book *The First Three Minutes*, written in 1977, remains the clearest summary of where cosmology stood at that time. Weinberg noted that progress is often delayed because scientists 'don't take their theories seriously enough': the history of the 'big bang' concept before 1965 (described in Chapter 3) certainly exemplifies this hesitancy. Cosmology's current buoyancy owes a lot to the incursion of particle physicists with Weinberg's robust intellectual confidence.

The electromagnetic and the 'weak' forces merge at very high energies: these two forces acquired distinctive identities only after our universe cooled below some critical temperatures. Energies in particle physics are measured in giga-electronvolts, GeV for short, and the critical energy for this unification is about 100 GeV. When our universe was 10^{-12} seconds old, it would have been so hot that this would have been the typical energy of all its constituent particles. Such energies have already been reached by big particle accelerators, and Salam and Weinberg's concept of a unified 'electroweak' force has been vindicated.

The next goal is to unify the electroweak force with the strong or nuclear forces – to develop a so-called grand unified theory (GUT) of all the forces governing the *micro*physical world. (These theories still do not include gravity – that is an even greater challenge.) A stumbling block is that the critical energy at which unification occurs, the energy that is 100 GeV for the Salam–Weinberg theory, is thought to be 10^{15} GeV for the grand unification – a trillion times higher than any accelerator can reach. These theories can therefore be tested only if they have distinctive consequences in our low-energy world. Such theories predict, for example, that protons would not last for ever: they would decay very slowly – less than one atom per year in a whole tonne of material. (In the very long run, if the universe goes on expanding, all atoms would eventually dissolve away.)

If we are emboldened to extrapolate the big-bang theory far enough we find that in the first 10^{-36} seconds, but only then, the particles would be so energetic that they would all be colliding at 10^{15} GeV. However, this cosmic accelerator shut down more than 10 billion years ago, so one can learn nothing from its activities unless the era that ended when the universe was 10^{-36} seconds old left some fossils behind, just as most of the helium in the universe is left behind from the first few minutes.

Our ignorance of the physics governing the initial instants of cosmic expansion parallels the ignorance of atomic structure that stymied 19th-century speculations about the power supply inside the Sun. But there is an important difference. Atoms and nuclei could be probed by experiments, whereas the early universe is *the only place* manifesting these ultra-high-energy phenomena. This offers cosmologists a relationship with their physicist colleagues that is symbiotic rather than parasitic – cosmology may offer a crucial empirical input into the most fundamental issues of physics.

WHY MATTER AND NOT ANTIMATTER?

The initial instants of the big bang provided a 'free' accelerator. Physicists would seize enthusiastically at even the most trifling vestige surviving from an ultra-energetic early phase. But its traces may actually be very conspicuous indeed: all the atoms in the universe are essentially a fossil from an early stage, perhaps even as early as 10^{-36} seconds.

There are about 10^{80} protons within the observable universe, but there do not seem to be so many antiprotons. Why this asymmetry? The simplest universe, one might imagine, would contain particles and antiparticles in equal numbers. It is fortunate that our universe didn't possess this particular kind of symmetry. If it did, all protons would have annihilated with antiprotons as it expanded and cooled: it would have ended up full of radiation, but containing no matter, no atoms and no galaxies.

Antiparticles can be created in the laboratory by crashing particles together at very high speeds (close to the speed of light); but they annihilate when they encounter ordinary particles, converting their energy (mc^2) into radiation. No antimatter exists 'in bulk' anywhere on the Earth. It can survive only if 'quarantined' from ordinary matter: otherwise it signals its presence by the intense gamma rays generated when it annihilates. Our entire Galaxy is matter rather than antimatter. It is being churned up and intermixed by the recycling processes involved in stellar births and deaths (see Chapter 1); had it started off half matter and half antimatter there would by now be nothing left. But, on much larger scales, we cannot be so sure. It would be hard to refute a claim that entire 'superclusters' of galaxies consisted alternately of matter and antimatter. So why is our universe (or at least a large expanse within it) biased in favour of one kind of matter?

If protons couldn't be created or destroyed without the same thing happening to an equal number of antiprotons, then the excess protons, all 10^{80} of them, would need to have been there right from the beginning. This seems an unnaturally large number to accept as simply a part of the 'initial conditions'.

Andrei Sakharov was the first to address this problem. Although more widely celebrated for other achievements, Sakharov made several contributions to cosmology; he was particularly interested in the exotic physics of the ultra-early universe. In 1967, he presciently spelled out three conditions that would need to be fulfilled if the universe was to end up containing more matter than antimatter. The **first** is an obvious one: the quantity called 'baryon number', the difference between the number of particles and antiparticles, cannot be exactly conserved. (This contrasts, for instance, with net electric charge: positive charges can never appear without *exactly* equal negative charges appearing as well.) **Second**, the cosmic expansion must be so fast that it prevents complete equilibrium from being established; otherwise every reaction would be exactly balanced by its inverse. The **third** condition might seem harder to fulfil: the relevant reactions must *not be exactly time-reversible*, so that they can 'sense' the direction of time singled out by the cosmic expansion.

One of the most cherished results in physics is that all reactions obey what is called 'TCP symmetry'. This means that any reaction would proceed in exactly the same way as a 'mirror' reaction in which the direction of time (T) was inverted, and the electric charge (C) was reversed, along with left and right, or parity (P). If overall TCP symmetry is indeed sacrosanct, then T-symmetry would be violated if (and only if) CP was as well. Two years before Sakharov published his ideas, Jim Cronin and Val Fitch found, to everyone's great surprise, that some unstable particles called K-mesons occasionally (in 0.2 per cent of cases) decay in a fashion that violates CP symmetry: this implies, assuming TCP symmetry, that some processes in the microworld 'know' about the direction of time.

Fitch and Cronin's process involved just the 'weak' force. No full theory relating the weak and strong forces existed back in 1967. In so-called grand unified theories, developed since the late 1970s, the time (T) asymmetry would exist not just for the weak forces, where it has been observed, but also for the 'strong' nuclear forces involved in the creation and annihilation of baryons. Sakharov's criteria then allow a slight preference for creating particles rather than their antiparticles. There might, for example, be some very heavy particles, called X, their antiparticles being denoted X^1. When the universe cooled below 10^{15} GeV, these would decay into quarks and antiquarks, the fundamental particles from which protons and antiprotons are made. If the decay products of each X^1 were the precise 'antis' of the decay products of each X, then the outcome would be symmetrical between quarks and antiquarks. However, an imbalance in the decays could lead to slightly more quarks than antiquarks, which could later be manifested as an excess of baryons over antibaryons.

For every 10^9 baryon–antibaryon pairs, there could be one extra baryon. As the universe cools, antibaryons would all annihilate with baryons, giving photons. But for every billion pairs that annihilated, one baryon would survive because it couldn't find a partner to annihilate with. The photons, now cooled to very low energies, constitute the three-degree background. There are indeed about a billion times more photons in our observable universe (10^{89}) than there are baryons

(10^{80}). All the atoms in our universe could result from a small fractional bias in favour of matter over antimatter, imposed as the universe first cooled below 10^{15} GeV.

Baryons (or their constitutent quarks) could readily be created or destroyed at the ultra-high energies prevailing when the universe was 10^{-36} seconds old. 'Baryon number' (the number of baryons minus the number of antibaryons) is *not* strictly conserved: this nonconservation was a prerequisite for the emergence of matter in preference to antimatter from the big bang. Even in our present universe, these 'nonconservation' processes would allow baryons to decay. But the decay rates are imperceptible: they would eliminate no more than one atom from each person's body during their life. (The decay time would have to be 10^{15} times faster for the radioactivity due to this effect to be a hazard.) Several experimental groups have built huge tanks of water (deep underground, to minimise the effects of cosmic rays and other disturbances), instrumented sensitively enough to record the decay of a hydrogen atom in a single H_2O molecule anywhere in the tank. They have had no luck yet.[1]

Grand unified theories are still tentative, but they at least bring a new set of questions – the origin of matter, for instance – into the scope of serious discussion.

Numbers like the ratio of the baryon density to the photon density are universal cosmic numbers, in the sense that they may take the same value throughout our observable universe; they are the outcome of microphysical processes (particle collisions and annihilations), which take place while the primordial material expands and cools through some critical temperature range. But this in a sense begs still more fundamental questions. Why, for instance, do the microphysical laws have the small inbuilt asymmetry that Sakharov's idea requires? The answer to that question may lie even further back than 10^{-36} seconds, and involve even higher energies than those of 'grand unification'.

GRAVITY AND UNIFICATION

The earliest phases of the big bang confront us with conditions so extreme that we know for sure that we do *not* know enough physics. The two great foundations of 20th-century physics are the quantum principle and Einstein's general relativity. The conceptual structures erected on these foundations are still disjoint: there is generally no overlap between their respective domains of relevance. Gravity is so weak that it's negligible on the scale of single molecules, where quantum effects are crucial. Celestial bodies whose motions are governed by gravity are so massive that quantum effects can be ignored. Heisenberg's uncertainty principle tells us that we cannot measure both the speed and the position of an atomic particle, but this intrinsic 'blurring' is negligible for a planet, star or galaxy. But what if we extrapolate back to the stage when everything we can now see, out to our 'horizon' 10 billion light-years away, was squeezed smaller than a single atom? At this stupendous density, attained in the first 10^{-43} seconds (known as the Planck time), quantum effects and gravity would both be important. What happens when quantum effects shake an entire universe?

Physics is incomplete and conceptually unsatisfactory in that we lack an adequate theory of *quantum gravity*. Some theorists believe it is no longer premature to explore what physical laws prevailed at the Planck time, and have already come up with fascinating ideas; there is no consensus, though, about which concepts might really 'fly'. We must certainly jettison cherished commonsense notions of space and time: space-time on this tiny scale may have a chaotic foamlike structure, with no well-defined arrow of time; there may be no timelike dimension at all; tiny black holes may be continually appearing and merging. The activity may be violent enough to spawn new domains of space-time which evolve into separate universes.

Later events (especially the 'inflation' phase described in the next chapter) may erase all traces of the initial quantum era. And the only other arena for quantum gravity effects is near the central singularity within black holes – whence no signals can escape. A theory that has no manifest consequences except in such exotic and inaccessible domains is hard to check. To be taken seriously it must either be rigidly embedded in some all-embracing theory that can be tested in many other ways, or else it must be perceived to have a unique inevitability about it – a resounding ring of truth that compels assent.

The approach that offers the best hope of unifying all the forces is 'superstrings'. According to this view of the world, the basic entities are strings rather than points or particles. Strings can vibrate in various harmonics, and different particles correspond to specific modes of vibration. The strings are tiny. They are about 10^{20} times smaller than an atomic nucleus: a string is smaller than a nucleus by about the same factor by which a nucleus is smaller than we are.

The most promising superstring theories require 10 dimensions. We don't notice these extra dimensions because they are 'compactified', rather as a sheet of paper, a two-dimensional surface, might look like a one-dimensional line if rolled up very tightly. Each 'point' in our familiar space has, in effect, six dimensions of internal structure. The appeal of superstring theory is that it may explain not only the basic particles but the properties of space as well. Einstein's general relativity, which interprets gravity as curvature in four-dimensional space-time, is inescapably built into the theory: the quantum of gravity, the graviton, is the simplest vibration mode of a superstring loop.

The current challenge is to understand why the 10-dimensional space 'compactifies' into our familiar four dimensions (time, plus three dimensions of space), rather than into three or five dimensions; and also to pick out the unique way in which this happened. There is still an unbridged gap between 10-dimensional superstring theory and observable phenomena. Superstring theory poses questions that are still too hard for mathematicians to answer. In this respect it is different from most physical theories: usually the appropriate mathematics has been developed beforehand. Einstein, for instance, used geometrical concepts developed in the 19th century, and didn't have to develop from scratch the mathematics he needed.

The physicist Eugene Wigner wrote a celebrated article on this theme entitled 'The unreasonable effectiveness of mathematics in the physical sciences'. It is indeed remarkable that the external world displays so many patterns that our minds can interpret in mathematical 'language' – especially when these are so remote from the everyday experiences and phenomena that our brains have evolved to cope with. Edward Witten, the leading 'superstring' theorist, describes the idea as '21st-century physics that has fallen into the 20th century'. But it will be remarkable if humans of any century can develop a theory as 'final' and comprehensive as superstrings are claimed to be.

COSMIC HISTORY, PARTS 1, 2 AND 3

The history of our universe divides into three parts:

1. The first millisecond, a brief but eventful era spanning 40 powers of 10 in time, starting at the Planck era (10^{-43} seconds). This is the intellectual habitat of mathematical physicists and quantum cosmologists. The relevant physics is still speculative – indeed, one motive for studying cosmology is that the early universe may offer the only real clues to the laws of nature at extreme energies.

2. The second stage runs from a millisecond to about a million years. It's an era where cautious empiricists feel more at home. The densities are far below nuclear density, but everything is still expanding quite smoothly. There is good quantitative evidence – the cosmic helium and deuterium abundances, the background radiation and so on, which were described in Chapter 3 – and the relevant physics is well tested in the lab. Part 2 of cosmic history, though it lies in the remote past, is the easiest to understand.

3. But the tractability lasts only so long as the universe remains amorphous and structureless. When the first gravitationally bound structures condense out – when the first stars, galaxies and quasars have formed and lit up – the era studied by traditional astronomers begins. We then witness complex manifestations of well-known basic laws. Part 3 of cosmic history is difficult for the same reason as all environmental sciences – from meteorology to ecology – are difficult: they involve ultra-complex manifestations of simple laws.

Cosmology confronts us with two contrasting styles of problem. The early evolution of our universe (Parts 1 and 2), when everything expands almost uniformly and no structures have yet condensed out, can be described by just a few numbers – just as, for instance, the physics of subatomic particles can. But the genesis of an individual galaxy like our Milky Way involves gas dynamics, star formation and the feedback from stars and supernovae. Understanding these complicated and 'messy' processes, the complexities we see around us and are part of, requires a different approach – as different, in the words of the relativity theorist Werner Israel, as mudwrestling is from chess.

Theories of how galaxies evolve will never be as 'clean' as the chesslike theories that particle physicists aspire to: the former will be more like a good theory in geophysics. For instance, continental drift (plate tectonics) is a unifying idea that gives insight into previously unrelated facts, but it is no disparagement of this theory that it cannot predict the exact shape of the continents.

HOW MUCH WOULD A FINAL THEORY MEAN?

The ultra-early universe may one day be triumphantly subsumed into some overarching theory that applies from the Planck time (10^{-43} seconds) onwards. Indeed some physicists already claim that our universe evolved essentially from 'nothing'. But they should watch their language, especially when talking to philosphers. The physicist's vacuum is a far richer construct than the philosopher's 'nothing': latent in it are all the particles and fields described by the equations of physics. In any case, such a claim doesn't bypass the philosophical question of why there *is* a universe. To quote Stephen Hawking, 'What is it that breathes fire into the equations? . . . Why does the Universe go to all the bother of existing?'

Nor would a fundamental theory help us to untangle the complexities of later cosmic evolution. We may be 'reductionists', believing that the complexities of chemistry and biology are in principle reducible to physics, and that even the most elaborate assemblages of atoms are governed by Schrödinger's equation. But that equation cannot in practice be solved for anything more complicated than a single molecule. The sciences are in a hierarchy of complexity, from particle physics, through chemistry and cell biology, to psychology and ecology. But each of these sciences is autonomous, in that it depends on its own set of concepts that can't be analysed into anything simpler.

To understand turbulent flow of water, a challenging and still unsolved problem, you must think in terms of wetness, whirls, eddies and so forth: analysing the water into atoms doesn't help – indeed it erases all its distinctive features. And – to take a different example – what goes on in a computer could be described in electrical terms, but that misses the essence, the logic encoded in those signals. We can't solve Schrödinger's equations for even the tiniest biological organism, but even if we could it would never yield an economical description or insight. The way complex systems behave may be reducible to physics, but their behaviour cannot be reconstructed from the equations of basic physics. The sciences are linked together, but not into a hierarchy whose entire superstructure is imperilled by still-uncertain foundations.

There is a sense, but a rather restricted one, in which some sciences can claim to be specially 'fundamental'. Causal chains – if you go on asking why? why? why? – lead back to a question in particle physics or cosmology. For that reason (as Steven Weinberg, in particular, has emphasised) progress in these subjects will surely disclose some deep aspects of reality. We pursue them for that reason; not because the rest of science depends on them.

The limitations are well portrayed by one of Richard Feynman's favourite metaphors. Suppose you were unfamiliar with the game of chess. Then, just by watching it being played, you could gradually infer the rules of the game. The physicist, likewise, finds patterns in the natural world, and learns what dynamics and transformations govern its basic elements. But, in chess, learning how the pieces move is just a trivial preliminary to the absorbing progression from novice to grand master. The whole fascination of that game lies in the variety implicit in a few simple rules. Likewise, all that's happened in the universe over the last 10 billion years – the emergence of galaxies and stars, and the intricate evolution, on a planet around at least one star, that's led to creatures able to wonder about it all – may be implicit in a few fundamental equations. But exploring this complexity offers an unending challenge that's barely begun.

NOTE

1 These experiments have not been completely 'null'. They achieved an uncovenanted bonus when, as described in Chapter 6, they detected neutrinos from the supernova 1987A. Other experiments in the same underground laboratories might also detect the particles making up the dark matter.

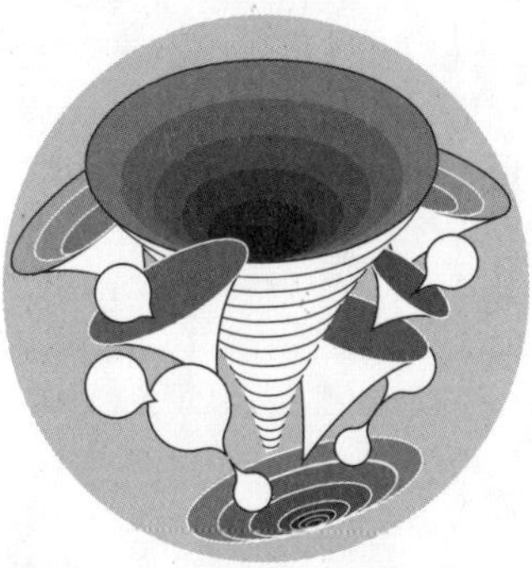

10

INFLATION AND THE MULTIVERSE

> And he showed me more, a little thing, the size of a hazelnut, on the palm of my hand, round like a ball. I looked at it thoughtfully and wondered, 'What is this?' And the answer came: 'It is all that is made'. I marvelled that it continued to exist and did not suddenly disintegrate; it was so small.
>
> Julian of Norwich
> (c. 1400)

THE 'FLATNESS' AND 'HORIZON' PROBLEMS

A universe may expand for ever; or it may eventually recollapse. Until we know what omega is, we won't know the eventual fate of our own universe. The two long-range forecasts – perpetual expansion, or recollapse to a big crunch – seem very different. But if we extrapolate backwards, we are faced with another puzzle: the starting points that could have led to anything like our present universe are actually very restrictive – very special – compared with the range of expanding universes that can be imagined.

Our universe is still expanding after 10 billion years. Other universes could have recollapsed sooner, not allowing enough time for stars to evolve: indeed, if a universe recollapsed within a million years, it would have never cooled below about 3000 degrees – for its entire life cycle, it would be an opaque fireball, with everything at the same temperature. Slightly *slower* initial expansion would have led to a very different universe from our own. And so would an expansion that was *too fast*: the expansion energy would then have overwhelmed gravity, and galaxies would never have been able to condense out. (Omega, though uncertain, definitely isn't enormously less than one.) In Newtonian terms the initial potential and kinetic energies must have been very closely matched. It's like sitting at the bottom of a well and throwing a stone up so that it just comes to a halt *exactly* at the top.

It is a basic mystery why our universe is still, after 10^{10} years, expanding with a value of omega not too different from unity. It looks surprising that it has neither collapsed long ago, nor is expanding so fast that its kinetic energy has overwhelmed the effect of gravity by many powers of 10. Our universe must have been given a very finely tuned impetus, exactly enough to balance the decelerating tendency of gravity, to have ended up as it is now. This is called the *flatness problem*.

The related *horizon problem* is even more perplexing. Why should the universe be expanding so uniformly and symetrically? A universe that was wildly inhomogeneous and anisotropic would seem to have more options open to it. So why were all parts synchronised to start expanding in this same special way, obeying the same dynamics? If remote regions started off differently, any drastic non-uniformities have somehow been ironed out.

At first sight, this might not seem a problem at all: when everything was closer together, shouldn't it have been easier for pressure waves and so forth to have homogenised our universe? But, actually, causal contact was *worse* when everything was more compressed. This is because, even though distances are shrunk in the early universe, the timescales are shrunk even more.

This communication problem in the earlier universe may be

clarified by an example, using actual numbers. Imagine a galaxy that now lies a billion light-years from us. Our universe has been expanding for 10–20 billion years, so there would be time for 10–20 signals to be exchanged during the present Hubble timescale. When our universe was 1000 times more compressed, and the background temperature, instead of being 2.7 degrees, was nearly 3000 degrees, this galaxy (it would then of course have been just a protogalaxy) would have been 1000 times closer – only a million light-years away from us, instead of a billion. If the galaxies were moving apart from each other at a *constant* speed, it would then have been 1000 times younger. The same number (10–20) of signals could be exchanged because, though each signal has 1000 times less far to go, the time available (the Hubble time) would be shorter by the same factor. But gravity is slowing the expansion down. When our universe was 1000 times more compressed, it was actually *more than 10 000 times* younger. So only one signal (and perhaps not even that) could have been exchanged at that early era.

Light signals could be exchanged *less easily* in the past than now; and nothing (no pressure wave, or other homogenising effect) can travel faster than light. So why, when we observe remote regions in opposite directions, do they look so similar and synchronised? Why is the temperature measured by the COBE satellite almost the same all over the sky?

AN EARLY ACCELERATING PHASE

The horizon problem arises because gravity *slows down* the cosmic expansion: when our universe was younger and more compressed, it was expanding much faster, allowing less time for transmission of signals or any casual contact. The problem would be solved if the very early universe experienced an *accelerating* phase of exponential expansion. In an accelerating universe, causal contact would have been better at earlier times, so remotely separated parts of our present universe could have synchronised and coordinated themselves very early on, and then accelerated apart.

According to the inflationary theory, the reason why our universe is so big, and why gravity and expansion are so closely balanced, lies in something remarkable that happened during the first 10^{-36} seconds, when our entire observable universe was the size of a golfball. Ever since that time, the cosmic expansion has been *decelerating*, because of the gravitational pull that each part of the universe exerts on everything else. But theoretical physicists have come up with serious (though still, of course, tentative) reasons why, at the colossal densities before that time, a new kind of 'cosmical repulsion' might come into play and overwhelm 'ordinary' gravity. Very early on, the expansion would have been exponentially *accelerated*, so that an embryo universe could have inflated, homogenised and established the 'fine-tuned' balance between gravitational and kinetic energy when it was only 10^{-36} seconds old.

The repulsion arises because space itself was very different in that initial era. Before the nuclear and electromagnetic forces had acquired their separate identities, empty space (what physicists call 'the vacuum') would have had a huge store of energy latent in it; but this form of energy had the seemingly perverse property that it made the pressure *negative* (in other words space had a tension).[1]

According to Einstein's equations, a positive vacuum energy would cause a 'cosmic repulsion': the universal expansion would accelerate. This is quite opposite to what happens when the same energy is in a more familiar form. According to the 'inflation' theory, the ultra-early universe went through a phase when the vacuum energy was enormous, and the cosmic expansion was consequently exceedingly rapid. The 'inflation' ended when the vacuum decayed into a more ordinary state. This transition releases heat, rather as water releases 'latent heat' when it freezes;[2] the heat survives, cooled and diluted, as the 2.7-degree background radiation.

This remarkable idea was advanced by Alan Guth, an American physicist whose first serious incursion into cosmology led to a 'spectacular realisation' (to quote his own description of his insight). Guth's achievement was to spell out clearly why there might have been an early phase of accelerating expansion, and how it could lead to a universe as vast and uniform as we see around us. There were several earlier 'premonitions' of this hypothesis. For instance, I remember a summer-school lecture by the Belgian theorist François Englert, in which he discussed an exponentially growing universe. I certainly didn't appreciate how radical and important his suggestion was, and am consoled by the fact that the rest of his audience seemed equally unresponsive. And there were other prescient papers, by Alexei Starobinsky in the Soviet Union, by Richard Gott in the US and by Katsuoko Sato in Japan.

The discovery of the microwave background – the most important observational development in cosmology since Hubble's work – was preceded by various independent papers that were unread or misunderstood (see Chapter 3); so it is, sometimes, for theoretical breakthroughs.[3]

Guth's original proposal for 'driving', and then ending, the inflation ran into various snags. Indeed the mechanisms are still speculative, because they depend on physics at ultra-high energies, which is almost completely unknown. But the generic idea remains compellingly attractive because it seems to solve the flatness and horizon problems. Indeed it suggests *why* the universe is expanding – something that seems otherwise just a part of the 'initial conditions'. Previously, the uniformity of the universe seemed a mystery, and no reason could be given why the universe had heaved itself up to its observed dimensions. But during the inflationary phase the universe receives an outward impetus sufficient to sustain its expansion. Indeed Guth found it more difficult to understand how the inflation could *stop* (this became known as the problem of the 'graceful exit' from inflation).

The idea of 'inflation' is now more than 15 years old. There is still no consensus on the link between any specific unified theory and the mechanics of the inflation. But cosmologists can infer something about the physical laws that prevailed in the first 10^{-36} seconds of our universe's history – at the very least, we can rule out many options that would lead to a present universe very different from ours.

It may seem counterintuitive that an entire universe at least 10 billion light-years across (and probably spreading far beyond our present horizon) can have emerged from an infinitesimal speck. What makes this possible is that, however much inflation occurs, the total net energy is zero. It is as though the universe were making for itself a 'gravitational pit' so deep that everything in it has a negative gravitational energy exactly equal to its rest-mass energy (mc^2). This realisation makes it easier to swallow the concept that our entire universe emerged almost *ex nihilo*.[4]

THE FLUCTUATIONS: WHAT IS *Q*?

Inflation can stretch a universe 'flat', and explain its vast scale. But can it also account for the 'magic number' *Q*, whose value 10^{-5} characterises the 'ripples' or fluctuations from which cosmic structures have formed?' (See chapter 7.) When inflation was still a novel concept, back in 1982, the leading theoretical pundits gathered for three weeks in Cambridge to debate and develop it. Discussion focused on the fluctuations: the outcome was initially frustrating, because the most 'natural' value of *Q* seemed to be about one, rather than as small as 10^{-5}.

The fluctuations from which clusters and superclusters form, and the even vaster ones spread right across the sky that COBE has now mapped, are the outcome of microscopic quantum processes at an ultra-ancient epoch when the universe was squeezed smaller than a golfball. We now understand how *Q* depends on the details of inflation: we can pick specific assumptions about the physics of inflation, work through the mathematics, and find out, for each assumption, what the ripples should look like. By comparing the outcome of these calculations with observations, we can at least narrow down the range of physical theories that are tenable. The actual number 10^{-5} still does not, however, have any natural explanation.

By mapping clusters and superclusters, and probing the microwave background, observers are confronting the inflationary era of cosmic expansion (10^{-36} seconds) with *real empirical tests*, just as we can already, by measuring the present abundances of helium and deuterium, learn about physical conditions during the first few seconds. There is an iteration between well-defined (albeit speculative) theories and data that can constrain them: to that extent, inflationary theories are squarely within the frame of serious science.

THE STATUS OF INFLATION

In the earliest variants of the inflationary universe theory (and the simplest to visualise) everything starts off in an initial simple 'bang'; the inflation is then an intermediate interlude which 'stretches' the universe enough to solve the so-called 'flatness problem'. This requires inflation by a factor of at least 10^{30}. However, the inflation factor is likely to be far bigger: a tiny initial region would then be stretched not merely to the presently observable horizon, but much larger still. Our universe would then be destined to continue expanding for much longer than it has already; its present density would consequently be very close to the 'critical' value which demarcates ever-expanding and recollapsing universes. Most versions of inflation theory, therefore, predict that the quantity cosmologists call omega should be *almost exactly* equal to one. Our universe extends vastly beyond our present 10–20-billion-light-year horizon; many more galaxies will come into view as it continues expanding. It may still eventually recollapse, but only after it has enlarged by a further factor $10^{1\,000\,000}$!

Most theorists, especially those who have come to cosmology from a background in high-energy particle physics, regard inflation as a beautiful idea, offering a compelling insight into why our universe has its distinctive properties. But some are less enraptured, particularly those who prefer a more geometrical approach. The most distinguished of these is Roger Penrose. For him, inflation is a 'fashion the high-energy physicists have visited on the cosmologists'; he notes that 'even aardvarks think their offspring are beautiful'. Despite such discordant voices, and some innovative ideas along alternative lines, most thinking about the ultra-early universe incorporates the concept of inflation.

Variants of the concept have spawned and mutated – and some of these variants suggest that universes can do the same. The Russian cosmologist Andrei Linde advocates *chaotic inflation* – a more complex scenario where the entire universe (the 'multiverse' in the terminology I'm using) could be infinite and eternal, but continually generates inflating regions which evolve into separate universes[5]. What we call our universe may be just one domain, or one episode, of an eternally reproducing cycle of different universes. These are now disconnected from ours, but can be traced back to common ancestors. The big bang that led to our universe is just one event in a grander structure. In the early 1970s, Sakharov proposed something along these lines: he called it the 'multi-sheeted' universe. But ideas have become more concrete in the context of 'inflationary' concepts.

TOWARDS OTHER UNIVERSES

All fundamental forces governing our universe – gravity, nuclear and electromagnetic – are different aspects of a single primeval force. Transitions in the properties of space itself, the 'vacuum', occurring as a universe cools down, differentiate the forces and establish the masses of the elementary particles. Such a transition probably terminated the early 'inflationary' expansion that smoothed out 'patches' large enough to become universes like our own.

These changes in the vacuum are like the 'phase' transitions that occur between gas, liquid and solid when ordinary materials are cooled. Their imprint may be arbitrary or 'accidental', like the patterns of ice on a pond. (Another analogy is with the way an ordinary magnet behaves: at high temperatures, magnetism disappears because the individual atoms are so thermally agitated that they are randomly oriented, but when a magnetic substance is cooled below a specific temperature, called the Curie point, the atoms spontaneously 'line up', but in a direction that is generally unpredictable.) Separate universes, or separate domains within an infinite universe, might have cooled down differently, even ending up governed by different laws.

Real space cannot be infinitely subdivided. Only 40 powers of 10 down the ladder from the terrestrial scale bring us to the Planck scale, the smallest length allowed by quantum uncertainty in the fabric of space. Our present Hubble radius, which sets the horizon for any present observations, is only 40 powers of 10 larger than atomic dimensions. But there is no *upper* bound to the dimensions that may eventually come into view: beyond our present Hubble radius could lie layer upon layer of larger structure. Our part of the universe may be fated to collapse after (say) $10^{100\,000}$ years – one followed by a hundred thousand zeros, rather than one followed by just ten zeros (its present age). The step from our present Hubble radius up to the overall scale of our universe may be much bigger than the step from a single particle to the Hubble radius. Light that reaches us in the far future, from regions far beyond our present horizon, may reveal that we occupy a (perhaps atypical) patch embedded in a grander structure. We could even, for instance, inhabit a finite or 'island' universe, whose edge may sometime come into view.

Even a universe that collapses, after tracing out a vast cosmic cycle, need in no sense be the whole of reality: in the grander perspective of the multiverse, it is just one 'episode', or one domain. An 'eternally inflating' multiverse may sprout separate domains; the laws of physics may vary between one universe and the next. Moreover, inside every black hole that collapses may lie the seeds of a new expanding universe.

The ensemble, the multiverse, could encompass universes governed by different laws and fundamental forces, and containing different kinds of particle. Universes would not live equally long, nor have equally eventful histories: some, like ours, may expand for much more than 10 billion years; others may be 'stillborn' because they recollapse after a brief existence, or because the physical laws governing them aren't rich enough to permit complex consequences. The 'ripple amplitude' Q in other universes may be much larger, or much smaller, than in ours. In some of them, space itself may even have different numbers of dimensions.

Only some universes (our own, of course, among them) would end up as propitious locations for complexity and evolution. This (literally infinitely) enlarged view of the cosmos is crucial for anthropic reasoning, discussed in Chapters 14 and 15. Other universes are not directly observable, but their conceptual status is on no worse a footing than superstrings (or even the more familiar quarks): these, too, are unobservable theoretical constructs whose manifestations help to account for the way the world is.

Our universe seems uniform simply because our present observational horizon is so small compared with the characteristic scales in the multiverse. But it may be 'special'. We are obviously not at a random point in space: we are on a planet warmed by a star. We are not surprised by this: we don't argue that it would be more 'typical' to be isolated in intergalactic space. Likewise, our 'home' universe has to be special – in its contents, and in the laws and forces governing it – for any life to evolve in it. But we can understand our 'home' universe better by realising that it is just one island in the cosmic archipelago.

Early cartographers speculated about continents beyond the frontiers of the then-known world, and about the dragons and serpents that populated terra incognita. Domains that we can't observe might seem to have a similarly fragile conceptual status. But their status is underpinned by some well-worked-out theories that at least constrain the 'dragons' that lie beyond our cosmic horizons.

NOTES

1 The underlying reason for the tension may be clarified by a very simple argument. Consider an empty jar (in other words, a jar containing just vacuum), sealed by a movable piston. Suppose the piston is pulled outward. The jar then contains more vacuum than before. If the vacuum has an energy, the contents of the jar have therefore *gained* energy. This is the opposite of what happens when the jar contains hot gas: the gas then gets diluted as it expands, and also *cools*; the heat-energy it loses is converted into work done by the pressure pushing the piston. So the vacuum must exert a 'negative pressure' on the piston. (Of course the analogy between space and a gas is incomplete. Space cannot be confined within a jar: only the expansion of an entire universe can actually change the amount of vacuum energy.)

2 The energy input from 'vacuum decay' could have made the universe hot enough for the processes discussed in Chapter 9 – for creating an excess of matter rather than antimatter – to come into play.

3 It is actually no coincidence that the idea of inflation caught on around 1980, and not much before. That was the time when 'grand unified' theories were being devised in which baryons were not exactly conserved: such theories gave substance to Sakharov's scheme for generating an excess of matter over antimatter. If baryons had been strictly conserved, the 10^{80} we now observe must have been there right from $t = 0$; it would be hard to see how everything could have inflated from an infinitesimal volume if that huge number of particles had always been there. The physics must allow the entire volume to be populated with baryons after inflation is over, without an equal number of antibaryons appearing at the same time.

4 Some quantities, unlike baryon number, actually are strictly conserved – electric charge, for instance. The total electric charge in our universe may indeed, now and always, be exactly zero.

5 Chaotic inflation is described in, for instance, A. Linde's book *Particle Physics and Inflationary Cosmology* (New York: Harwood, 1990).

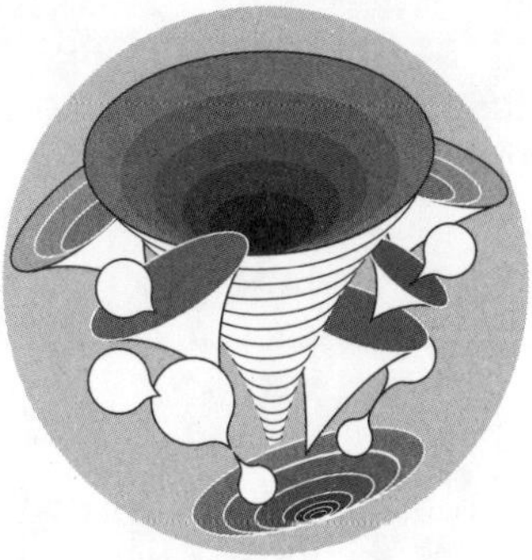

11

EXOTIC RELICS AND MISSING LINKS

I know of *nothing but* miracles.

Walt Whitman

IT IS ONLY BECAUSE WE PERCEIVE patterns and regularities in the natural world that science doesn't 'clog up' as data accumulate. On the contrary, as we come to see how previously disconnected facts hang together, and subsume data into more and more general laws, we need to remember *fewer* independent basic facts, from which all the rest can be deduced. We need not record the fall of every apple.

Physics and astronomy are the sciences that have, up till now, succeeded best in reducing the bewildering complexities of the natural world to just a few underlying principles. The regular courses of the moon and planets have been known since ancient times. We owe to Newton the great unifying idea that these motions are governed by the same gravitational force that holds us down on the Earth. The 19th-century Russian chemist Dmitri Mendeleev found patterns, in the properties of the 90-odd chemical elements. These patterns, manifest in the 'periodic table', we now attribute to the fact that atoms are made from just three basic constitutents: the protons and neutrons (baryons) making up the atomic nucleus, and the negatively charged electrons that orbit the nucleus according to the laws of quantum mechanics. Over the history of our Galaxy, all the atoms of the periodic table have been built up from the two simplest: hydrogen and helium. These two are themselves the outcome of nuclear reactions in the first few minutes of the big bang.

The lure of the fundamental is very powerful. The great insights in science come from 'systematising' or unifying phenomena that previously didn't fall into a natural pattern. All physicists aspire to unify the forces of nature, or develop some grand synthesis of Einstein's relativity and quantum theory. These are plainly 'number-one problems'. But an undue focus of talent in one highly theoretical area can be frustrating for all but a very few exceptional (or lucky) individuals.

In deciding where to dedicate one's individual scientific effort, it isn't necessarily sensible to aim straight for the most important problem, which may be neither timely nor tractable. A better strategy is to maximise the importance of the problem, *multiplied by* one's chance of solving it. In a famous essay, Peter Medawar reminds us that 'No scientist is admired for failing in the attempt to solve problems that lie beyond his competence. The most he can hope for is the kindly contempt earned by Utopian politicians. If politics is the art of the possible, research is surely the art of the soluble. . . . Good scientists study the most important problems they think they can solve.'

Scientific effort is deployed in a very patchy way. Some fashionable fields are intensively cultivated: whenever one of the 'leaders of the pack' proposes a new idea, it is quickly seized on and pushed forward by a phalanx of talented (usually younger) theorists. But in other fields interesting and timely projects get little attention, simply because those with the relevant expertise are already engrossed in something else.

Searches for unified theories have come a long way since

Mendeleev's time. The goal is to find underlying patterns and relationships among particles and forces, and thereby render the subatomic world less confusing and arbitrary-seeming. In the 1970s physicists developed what came to be called the 'standard model'. This brought some order into the study of the subnuclear world of quarks, electrons and other particles, but the number of 'elementary particles' remains depressingly large, and the equations still involve 18 numbers that have to be determined by experiment, and can't be derived or interrelated by the theory. Nothing better is likely to be achieved without some new experimental discoveries. As the particle physicist John Polkinghorne has written, 'Given the restless and competitive spirits of bright young theorists, ever anxious to gain a reputation by refuting or replacing the ideas of their elders, it is hard to believe that there are many, or indeed any, rational alternatives which have escaped notice through a socially induced slothful acquiescence in the status quo.'

The 'standard model' was a real advance – it accounts for the results of most experiments, even with high-energy accelerators. But it's plainly unsatisfactory that so many separate numbers remain unexplained. There is little consensus on the next step towards 'unification' of the basic physical forces. One of the stumbling blocks is that the distinctive features of unified theories probably manifest themselves at energies that occurred only in the initial instants of our universe – nowhere else at all, except maybe deep inside black holes. Mathematically elaborate theories such as superstrings are still disjoint from any real experiments.

Taken to excess, the tendency towards 'baroque' theorising can be unhealthy: the essence of any science is confrontation with experiment and observations, not retreat from them. Hence the importance of the interface with cosmology: very near the beginning of our universe, particles would have been moving around with far more energy than can be generated in any laboratory experiments.

Theorists are exploring a whole range of ideas on how the forces and particles may be unified: with luck, one may be correct. Can cosmology offer any clues to which one? Do these theories, for instance, predict anything about the present universe that is manifestly incompatible with what we see? Or (better still) do they predict any relics or fossils that might actually be discovered?

Using well-established physics we can extrapolate cosmic evolution back to the stage when the universe was a millisecond old

(10^{-3} seconds); the most powerful particle accelerators can generate the conditions that prevailed at 10^{-14} seconds; earlier than that, energies would have been higher still. But many crucial features of our universe could have been imprinted when the cosmic clock was reading 10^{-36} seconds, or even earlier. In these contexts, each factor of 10 in the age of the universe – each extra zero on the reading of the cosmic clock – should be counted equally. The leap back from 10^{-12} seconds to 10^{-36} seconds is then bigger (in that it spans more factors of 10) than the timespan between the three minutes when helium was formed (about 200 seconds), and the present time (about 3.10^{17} seconds, or 10 billion years).

The quest for 'fossils' of that ultra-early era, 'missing links' between the cosmos and the microworld, is as important for cosmologists as for particle physicists. Three extraordinary possibilities, any of which might one day be discovered, are described in this chapter: black holes as massive as a mountain, but smaller than a single atom; magnetic monopoles; and strings stretching across the universe, thinner than an atomic nucleus but so heavy that their gravity could have triggered galaxy formation.

HAWKING RADIATION: A UNIFYING IDEA

By 1974, Stephen Hawking was already acclaimed for his gravitational researches. He had led the resurgence in the subject that stemmed from Penrose's mathematical insights, and had elucidated the nature of black holes. The idea that black holes actually existed was starting to be taken seriously.

But he was becoming increasingly frail. On many days, I wheeled him into his office. He couldn't lift a book, or even turn its pages. I would open a text on quantum electrodynamics – the theory developed by Richard Feynman, Freeman Dyson and others, which accounted with fantastic accuracy for the radiation from electrons and atoms. Each day he sat – hunched, almost motionless – for several hours, reading and thinking. His illness seemed to be overwhelming him; I held little hope that he would achieve anything further. But after these months of cerebration he achieved a new insight that connected the Feynman and Dyson theories (which had nothing to do with gravity) with Einstein's relativity. He showed that black holes were not completely black, but would actually radiate; as a bonus, he found new connections between gravity and thermodynamics. Dyson rated this conceptual breakthrough as 'one of the great unifying ideas in physics'.

Hawking has continued unflagging in his research, especially on the overarching problem of reconciling quantum theory with cosmology. But the latter issues remain controversial: 20 years later several contending 'schools' still espouse alternative viewpoints, and it is unclear whether Hawking's favoured approach will prevail. But the significance of his 1974–75 papers on the quantum radiance of black holes has been generally acknowledged. The ideas were, however, so contrary to earlier intuitions that they proved hard to assimilate at the time. When Hawking first presented his idea, at a conference in Oxford, the session chairman, John Taylor, was overtly disparaging: soon afterwards, Taylor, together with Paul Davies (later to become a distinguished populariser of science) published a paper 'refuting' the idea. It took more than a year even for some leading experts – people such as Zeldovich and his colleagues in Moscow – to become convinced of Hawking's claims. The concept of black-hole evaporation, whether or not the predicted effects can be observed, ranks as a pinnacle in our understanding of gravity.

MINIHOLES

Black holes can have any size: their radius is proportional to their mass. The huge holes in the centres of galaxies, as heavy as a billion suns, could engulf our entire Solar System. A hole weighing as much as the Sun would be six kilometres across; a hole's diameter would be only nine millimetres if it had the mass of the Earth. What about black holes the size of atoms? Such holes would still be enormously *heavier* than atoms. A hole weighing a million tonnes would fit inside an atomic nucleus. To make it, about 10^{36} protons would have to be packed within the space normally occupied by one. This particular large number should come as no surprise. The electrical repulsion between two protons is 10^{36} times stronger than the gravitational attraction between them. This same huge number of protons must therefore be packed together before gravity can compete with the vastly stronger electric and nuclear forces.

Anything orbiting close to a black hole feels a stronger gravitational pull on the side facing the hole than on the side that is further away. The difference between these two pulls, the 'tidal' force, tears apart any star (or any planet or astronaut likewise) that comes too close. The gravitational and tidal forces are fiercer around miniature atomic-scale black holes – so much so that they affect even individual electrons and protons.

Empty space, 'the vacuum', seethes with activity on the microscopic quantum scale. Heisenberg's famous uncertainty relation allows particles to 'borrow' energy for a very short time: a particle and its antiparticle can briefly borrow their entire rest-mass energy (mc^2), and thereby have a transient existence. These 'virtual' pairs of particles are latent everywhere. But, near a small black hole, gravity can accelerate a particle so violently that it recoups energy mc^2 within the brief span of time for which the uncertainty relation allows this energy to be borrowed. It need not then annihilate to repay the debt: a virtual pair can instead transform into a *real* particle and a *real* antiparticle. One escapes; the other is trapped inside the hole with, in effect, negative energy. The hole shrinks slightly (it has gained 'negative energy'); the other particle, carrying positive energy, escapes the hole's clutches.

So black holes actually emit something: they glow, rather than being completely 'black'. But holes can emit (for instance) electrons only if they are themselves no bigger than an electron – this effect is only, therefore, important for 'miniholes'.[1] Larger holes are cooler; they cannot create electrons and positrons, but can radiate light or microwaves (or other kinds of radiation) with wavelengths larger than the size of the hole. The black holes that form when massive stars die would be only a millionth of a degree above absolute zero. For these (and, a fortiori, for the even bigger black holes in galactic centres), Hawking's process is utterly negligible: big holes radiate energy far more slowly than they soak it up from the 2.7-degree radiation that pervades even intergalactic space.

But the temperature of a 'minihole' the size of an atomic nucleus is a billion degrees. Intense radiation would be detectable if such an object came anywhere near the Earth. As a minihole radiates, it shrinks, and its emission gets hotter and still more intense, until it finally disappears in a flash of gamma rays.

Before Hawking reached these conclusions, several people had applied the concepts of heat and temperature to black holes. It was known that the area of the horizon around a black hole could increase (if, for instance, things fell into the hole and it gained mass) but could never decrease; if two holes merged, the horizon around the resultant hole would have a bigger area than the two original holes added together. This area inexorably increased; in that respect, it was like entropy – the quantity that measures 'disorder' or randomness. According to the famous second law of thermodynamics, entropy can never decrease.

The Israeli physicist Jacob Bekenstein (then one of John Wheeler's students at Princeton) pushed this resemblance further, pointing out that, if the area of a black hole resembled entropy, then the strength of gravity on its horizon resembled temperature. But he was uneasy because black holes were then presumed to be absolutely black, absorbing radiation but incapable of emitting any. Bekenstein's insight was vindicated when Hawking showed that they actually do radiate with a temperature that depends on the strength of gravity at their horizon.[2]

The theory of black-hole evaporation, sometimes called the 'quantum radiance' of black holes, has been 'battle-tested' through being rederived in alternative ways by different people. But that is no substitute for actually observing the predicted radiation. So could very small black holes actually exist?

EVAPORATING BLACK HOLES?

The universe is punctured by black holes, each marking the death of a star, or else the outcome of some runaway catastrophe in the centre of a galaxy. But no astronomical processes could create a black hole smaller than two or three solar masses. Any star of lower mass, even if it had exhausted all nuclear fuel, could survive indefinitely, quite stably, as a neutron star or white dwarf. It could transform into a black hole only if a huge external pressure squeezed or imploded it to even higher densities than a neutron star. Anything smaller than a star – a planet or asteroid – would need to be squeezed to a higher density still.

The only black holes that radiate significantly are very tiny ones. Creation of such holes by 'implosion' of a suitable mass was envisaged by Wheeler in the 1960s. Renewed interest, though still in a 'science-fiction' spirit, stems from the idea that space inside the hole, rather than being crushed indefinitely, could, through a quantum effect, sprout into an entire new universe. Alan Guth has written semi-facetiously about 'creating a universe in the laboratory' by imploding just 100 kilograms of material to such extreme densities that it became a tiny black hole. If we were to find a minihole, however, we need not interpret it as an artefact of a super-civilisation: it could be a relic of the early universe.

In its very early stages, our universe would have been squeezed far denser than a neutron star, and the tremendous pressures could, as Igor Novikov first realised, have created black holes much smaller than those that are forming in our present universe. The likelihood depends on how chaotic or irregular conditions were. It is not sufficient that the pressure should be high: pressure *differences* between one place and another are necessary to drive an implosion. The best guess about these fluctuations suggests (Chapter 7) that the pressure would have been too smooth. Minihole formation would require the early universe to have been rougher (larger-amplitude ripples or curvature fluctuations) on small scales than it was on the larger scales relevant to galaxy formation – like an ocean where the short waves are higher than the slow swell.

A still more extreme speculation is that the 'seeds' for black holes might form right back at the Planck time, from the 'space-time foam' itself; they would then be far smaller even than a single atomic nucleus, but could grow by accreting from their ultradense surroundings.

Black holes are possible candidates for the dark matter in galaxies, as described in Chapter 6. But very tiny holes are not dark – they are hot and radiate brightly. Holes weighing less than a trillion tonnes (sizes less than 10^{-10} cm) would not be 'dark' enough: there could not be enough to constitute all the dark matter without their combined radiation making more X-rays and gamma rays than are observed. But these miniholes could still, of course, exist in smaller numbers.

Holes with masses far below 10^{15} gm would not have survived from the big bang to the present epoch: they would have 'evaporated' long ago. Some would just now be in their final throes of evaporation.

Each of these miniholes, the size of a single proton, could radiate 10 gigawatts for the entire lifetime of the Solar System. These extraordinary entities are hypothetical, but their properties have been rigorously deduced from the two best-established theories in 20th-century physics – quantum electrodynamics, combined with general relativity. They are possible 'fossils' of the ultra-early universe – 'missing links' that might illuminate the quest for deep connections between gravity and the forces governing the microworld.

In their death-throes, miniholes emit gamma rays intense enough to reveal themselves several light-years away. In fact there may be an even more sensitive way of detecting them. The final explosion may eject an intense fireball of electrons and positrons. The weak magnetic fields that pervade all space interact with this fireball, converting its energy into a radio flash. Radio telescopes are enormously more sensitive than gamma-ray telescopes, and could detect such an event, caused by an entity smaller than an atomic nucleus, even if it happened in the Andromeda Galaxy – two million light years away.[3]

MONOPOLES

Michael Faraday, early in the 19th century, showed how magnetic and electric forces were interconnected: a moving magnet creates an electric force (and affects a galvanometer); conversely, movement of electric charges (an electric current) creates a magnetic field. These famous discoveries are the basis of dynamos and electric motors. Later in the 19th century, James Clerk Maxwell developed a theory that unified electricity and magnetism into 'electromagnetism'. In his equations, there is almost complete symmetry between the way the two forces appear – fast-changing electric fields cause magnetic fields, and vice versa. But there is one significant difference: positive and negative electric charges exist, but we cannot isolate a 'north' or a 'south' magnetic pole – chopping up a magnet just produces smaller magnets, each with two poles.

The basic carriers of electric charges are, of course, electrons. These have a 'standard' negative charge; protons have an exactly equal positive charge. In 1931, Paul Dirac was trying to understand why electric charges were quantised in this way. He devised an elegant explanation, which would work only if monopoles indeed existed; he even calculated the 'magnetic charge' that these hypothetical objects would need to possess if his theory was to be right.

Dirac did not expect monopoles to be especially heavy, and the failure to observe even a single one led to his idea being sidelined. The quest for monopoles has been a discouraging one. Straightforward arguments preclude there being too many of them: just as electric charges or conductors can 'short out' an electric field, magnetic monopoles (half with a 'north' magnetic charge, and half with 'south') would tend to cancel out a magnetic field if they were released into it. A magnetic field pervades our entire Galaxy, and it would not persist if there were too many monopoles.

Monopoles are now perceived in a new guise, as 'knots' in the vacuum. In the very early universe the vacuum itself would have been different. No atomic nuclei then existed: indeed the two forces that hold nuclei together – electric forces and the so-called 'strong' nuclear force – had no separate existence. These forces only acquired their distinctive character after a transition when the vacuum itself changed its nature. Before that transition, there was energy in the vacuum itself.[4]

Just as ice may not become a perfect crystal, so this cosmological transition leaves defects 'frozen' into space. The simplest such defects are magnetic monopoles. Two theorists, Andrei Polyakov and Gerard t'Hooft, independently realised that monopoles *must* have been created only 10^{-36} seconds after the 'big bang', when the forces differentiated. These monopoles, unlike those that Dirac envisaged, would be very massive – about 10^{15} times heavier than 'ordinary' particles – and would therefore be impossible to make in the lab. However, the number expected to have survived from the early universe seemed embarrassingly large: there would have been enough to short out the galactic magnetic field; even worse, their total mass would far exceed that of everything else in the universe (far too much, even, for the dark matter).

One of the successes that Guth claimed for his 'inflationary universe' was that it solved this so-called 'monopole problem': monopoles would be exponentially diluted during the inflation, to such an extent that there would be little chance of even one in our entire Galaxy. Sceptics about exotic physics might not be hugely impressed by a theoretical argument to explain the absence of particles that are themselves only hypothetical. Preventive medicine can readily seem 100 per cent effective against a disease that doesn't exist!

According to Polyakov and t'Hooft, monopoles have a very interesting structure. Their cores would, in effect, be a tiny sample of what the universe was like when it was 10^{-36} seconds old; surrounding the core would be a series of shells or layers mirroring conditions the universe passed through as it cooled. Each monopole would thus be a microscopic fossil of the uncertain early phases of cosmic history. A particle approaching a monopole head-on would experience, speeded-up and in reverse order, the conditions that prevailed in the big bang right back to 10^{-36} seconds.

In our present universe, a particle cannot be created (or annihilated) without the same thing happening to an antiparticle. But this law must have been violated in the early universe, as Sakharov first emphasised (see Chapter 9), to establish a preponderance of matter over antimatter – otherwise every proton would by now have annihilated with an antiproton, and our universe would just contain radiation. A proton that encounters the core of a monopole would re-enter an exotic environment like the ultra-early universe, in which it could annihilate, converting its mass into energy, even in the absence of its antiparticle.

Because the core of a monopole is so small, annihilations would be rare except where matter is packed very densely. The densest places in our present universe are the insides of neutron stars, where the mass of a mountain is squeezed into each cubic centimetre. (Even this is not nearly high enough to create a minihole, which would require further squeezing down to the size of a single atomic nucleus.) Neutron stars are the optimum location for annihilations for a second reason: as well as being the densest objects we know, they are the most strongly magnetised.[5]

Monopoles would 'home in' towards these strong magnetic fields. Those captured by a neutron star would settle towards its centre. Whenever one of these monopoles annihilated a particle, heat would be generated. There are about a hundred million neutron stars in our Galaxy, each a remnant of a supernova explosion. Most of these must be relatively 'cold': otherwise X-ray telescopes would have detected them. Not many monopoles, therefore, can have accumulated inside them. The coldness of old neutron stars actually constrains the number of monopoles in our Galaxy more tightly than the persistence of large-scale magnetic fields. The experimental quest for monopoles is a daunting challenge – they may exist, but astronomers are already sure that they are very sparse.

STRINGS

Monopoles are the simplest kind of 'knot', or 'topological defect', that could be trapped in space. A more dramatic possibility is that long linear defects called 'cosmic strings' could survive from the very early universe.[6]

Tom Kibble at the University of London pioneered the key ideas about cosmic strings back in the 1970s. When the vacuum changed from a high-energy state that drives inflation into what we now call 'empty space', some of the original vacuum state could get trapped, when our universe was 10^{-36} seconds old, inside very thin tubes. These would be 20 powers of 10 thinner than an atom, but would carry an immense mass of 10^{17} tonnes per metre.

Strings cannot have ends. Either they stretch right across our universe or they form closed loops, resembling rubber bands. The vibrations of an ordinary elastic string depend on its tension, and on how heavy it is (or how much inertia it has). Cosmic strings, though heavy, have such high tension that they vibrate at close to the speed of light. A lattice of 'open' strings would flail around. Sometimes two would intersect; or an open string could even tangle up and intersect itself. When the latter happens, a loop would pinch off. So the string network, at any stage, would consist of 'open' strings, stretching with the expanding universe, together with loops of different sizes.[7]

These cosmic strings should not be confused with the 'superstrings' mentioned in Chapter 9. The latter are hypothetical entities in 10-dimensional space which may underlie all the particles and forces of nature. Cosmic strings are also conjectural, in that their existence depends on uncertain features of the ultra-early universe. But, if they exist, they are objects of cosmic (rather than microscopic) length, and could have observable (even spectacular) effects in our present universe.

Even if the early universe were otherwise utterly smooth, the gravitational pull of the strings would create fluctuations. In the 1980s, Kibble, Alex Vilenkin and Neil Turok explored whether string loops could be the 'seeds' around which galaxies formed. The loops, with a hierarchy of scales, would be clustered in a fashion resembling that of galaxies. This particular idea, however, did not survive closer scrutiny. One problem, for instance, is that the string loops would have been hurtling through space at about a tenth the speed of light – so fast that their gravity could not have captured the surrounding material efficiently enough to build up a galaxy.

Though disappointing, this conclusion doesn't diminish the likelihood that strings exist – it simply means that the early universe must possess irregularities other than the strings themselves. Nor does it diminish the incentive to search for objects that straddle the microworld and the cosmic scale in such extraordinary ways.

What would happen if there were a string near us? A straight string distorts space so that a circle drawn around it closes up in slightly less than 360 degrees. If it were not moving, we should notice very little, and feel no gravity. But if it were moving, and sliced through us, then the distortion of space would manifest itself dramatically as our two halves squashed into each other at supersonic speed. Any realistic network of heavy strings would fortunately be so tenuous that there may be no string anywhere in our Galaxy, still less in our Solar System. So we must seek less dramatic effects.

Because strings distort the space around them, they deflect light and induce a distinctive kind of gravitational lensing. A background galaxy lying behind a long string would appear as two equal images, one on each side of the string. Searches for apparent pairs of galaxies answering this description have so far drawn a blank.[8] This is not, however, a fatal setback to the string hypothesis, because there wouldn't be enough strings to affect more than one part in 100 000 of the sky: astronomers would have needed remarkable luck to have pinpointed a string already. Any unambiguous evidence for gravitational lensing by a string would pin down how many tonnes per metre the string weighed, a quantity related directly to the details of unified theories, which are not yet pinned down by experiment. (That is of course another reason why physicists would like to find strings.)

Strings could reveal their presence indirectly, because their flailing motions generate gravitational radiation – waves of changing gravity that move through space at the speed of light. These waves could be detected as a byproduct of a quite unrelated research programme – accurate timing of pulsars.

The gravitational waves from strings would have periods of years, or even (for large loops) thousands of years. These waves would slightly distort all of space. It would be as though every star, in every galaxy, were jittering around its average position. The Sun's speed relative to any distant star would go up and down with every cycle of each such wave. If a star carried a clock transmitting regular 'beeps', the jitter in its distance would affect the time at which the signals arrived. For example, gravitational waves with a ten-year period would lead to the 'beeps' arriving earlier than average for five years, later than average for the next five years, and so forth. A jitter that amounted to (say) one kilometre would advance or delay the arrival by the time light takes to go a kilometre, about 3.3 microseconds. So if space were pervaded by gravitational waves, then the time kept by the beeping clocks would seem slightly erratic.

Remarkably, some stars do, in effect, carry precise clocks. These are the pulsars – spinning neutron stars with some kind of 'lighthouse beam' on their surface, from which we receive a radio beep once per revolution. Some pulsars, especially those that spin fastest (up to 600 revs per second) keep time very precisely indeed: the pulses are so short and sharp that their arrival times can be measured to a fraction of a microsecond. Even a motion amounting to 100 metres in a few years – three millimetres per hour – shows up in the timing data. If strings existed and were heavy enough for their gravitational fields to have induced galaxy formation, their gravitational waves would induce effects just about as big as this.

The concept that cosmic strings could seed galaxy formation is in itself extraordinary. But it is still more astonishing that these strings generate gravitational waves that Joseph Taylor and his colleagues can detect by observing irregular motions in neutron stars thousands of light-years away, even though the waves push the star slower than the *hour* hand of a watch!

Strings heavy enough to have influenced galaxy formation will soon either be discovered, or else will be ruled out. But it will be harder to detect or exclude lighter strings, so the quest will continue for these and other exotic relics of the eventful 10^{-35} seconds with which cosmic history began.

NOTES

1 A black hole has a well-defined temperature, proportional to the gravitational force just outside the hole. This force depends on M/r^2. The radius r of a hole scales with its mass M, so this 'Hawking temperature' depends on M divided by M^2 – it scales inversely with the hole's mass.

2 Whenever entropy increases, information is lost. For example, if we start off with a box partitioned into two parts, one containing cold gas (slow-moving atoms) and the other containing hot gas (fast-moving atoms), and then remove the partition so that the slow and fast atoms mix up, we know less about the location of each atom. When a black hole forms, all trace is lost of what it was made from. Its 'entropy' can be thought of as a measure of how many different possible ways it could have formed.

3 There is still debate about whether an evaporating hole would disappear completely, or leave some kind of relic. The answer depends partly on whether it carries some kind of conserved 'charge' which it cannot radiate away or neutralise. Electric charge is exactly conserved in the universe; nevertheless a hole that formed with such a charge could cancel it out by accreting particles with the opposite charge. On the other hand, a hole may not be able to get rid of the 'magnetic charge' it formed with; a remnant magnetically charged particle of mass 10^{-5} gm could then survive. Such particles should be added to the list of 'exotic relics' discussed in this chapter.

4 There would then have been, in effect, a huge 'cosmic repulsion': the universe would have expanded at an accelerating rate. This is the 'inflation' stage, described in Chapter 10.

5 Neutron stars are the dense remnants of supernova explosions; those that are still young are detected by astronomers as pulsars or X-ray sources (see Chapter 4).

6 Monopoles can be thought of as zero-dimensional defects (they are point-like); strings are one-dimensional. *Two*-dimensional defects, known as domain walls, could in principle occur as well, but these would be even more of an embarrassment than an excess of monopoles. Strings, on the other hand, would be welcome to many cosmologists.

7 The statistics of the networks – the sizes of the loops, and whether there would be more total length in open strings or in loops – are still controversial; it is hard to simulate a large enough part of the universe, for a long enough time, to be confident that the result is reliable. Flailing around at nearly the speed of light, strings emit gravitational waves. These waves carry away energy from the loops, which consequently shrink, eventually to tiny dimensions. (The final state – whether they disappear completely or leave some massive particle – is uncertain, just as it is for evaporating black holes.)

8 Astronomers have also looked for the distinctive effects that a fast-moving string would create in the microwave background. If a string moved across the sky, it would induce a redshift (and consequent apparent 'cooling' of the background radiation) on one side of itself, and a 'warming' on the other.

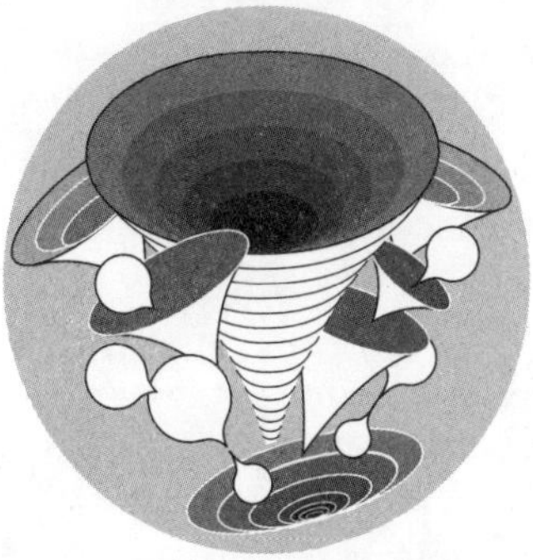

12

TOWARDS INFINITY: THE FAR FUTURE

In this great Celestial Creation, the Catastrophy of a World, such as ours, or even the total Dissolution of a System of Worlds, may possibly be no more to the great Author of Nature, than the most common Accident in Life with us, and in all Probability such final and general DoomsDays may be as frequent there, as even Birth-Days, or Mortality with us upon the Earth.

This Idea has something so chearful in it, that I owe I can never look upon the Stars without wondering why the whole World does not become Astronomers ... and reconcile them to all those little Difficulties incident to human Nature, without the least Anxiety ...

Thomas Wright of Durham
(1752)

THE NEXT HUNDRED BILLION YEARS

In about 5 billion years the Sun will die, swelling up into a red giant, engulfing the inner planets, and vaporising all life on Earth; it will then settle down as a slowly fading white dwarf. At about the same time (give or take a billion years) the Andromeda Galaxy, already falling towards us, will merge with our own Milky Way. When two galaxies merge, most of the constituent stars are unscathed. The chance of a head-on collision between individual stars is only one part in 100 billion. But the motions of all the stars would be severely perturbed, and the galactic 'grand design' of disc and spiral arms would be shattered.[1] The outcome would be a single amorphous swarm of stars resembling a bloated 'elliptical' galaxy.

What might our entire universe be like when it is ten times *older* than today – after, say, another 100 billion years? Cosmic expansion is being slowed by the gravitational pull that each galaxy exerts on everything else. If the density doesn't exceed the 'critical' value – if omega is one or less – our universe is fated to continue expanding. But if the density were substantially higher, gravity would have decelerated the expansion enough to bring it to a halt; indeed, if omega were as large as two, everything would, before 100 billion years have passed, have been engulfed in a final 'crunch'.

Most cosmologists would bet that our universe will still be expanding 100 billion years from now. They now believe that omega is either around 0.2 or else, for theoretical reasons, would guess that omega is almost exactly one. A show of hands among the participants at a conference in 1995 showed a roughly equal split between these views – though, reassuringly, only a small minority thought this a good way to settle scientific issues!

Theoretical prejudices change. In the early 1970s, observational clues were vaguer than they are now (and, such as they were, suggested that omega was small); but there was nevertheless quite widespread prejudice in favour of a finite 'closed' universe. Such a universe would collapse after a finite time, and would contain only a finite amount of material – the larger it was, the longer the 'cycle' would take. There seemed no particular reason why our universe should be enormously larger than the part we'd already seen. Nor why it should heave itself up to much more than its present size, or go on expanding for much longer than it has done already.

There were also some more 'philosophical' arguments for a closed universe. One was based on the so-called 'problem of inertia'. Galileo recognised that inside a windowless laboratory there is no way of knowing your speed – only your acceleration. But *rotation* seems different. When water in a bucket is spun, its surface deforms – it is depressed at the centre and rises towards the edge. Newton noted that this does not depend on whether the bucket itself shares the spin of its contents – the water's surface responds to rotation in some 'absolute frame', certainly not relative to the bucket.

Newton gave us the concept of an 'inertial' frame of reference – a spinning liquid, a gyroscope or a pendulum picks out a special frame, and we can test whether our laboratory is spinning relative to it. But it is still mysterious what determines this special non-rotating frame. Even before Newton's time, this was a philosophical issue. Among those who discussed it was the early 14th-century French philosopher Jean Buridan (famous for his uncompromisingly logical 'ass' which starved indecisively midway between two equally succulent bales of hay): 'To celestial bodies ought to be attributed the nobler conditions. . . . But it is nobler and more perfect to be at rest than to be moved. Therefore the highest sphere ought to be at rest.'

The 19th-century Austrian physicist and philosopher Ernst Mach argued that the inertial frame was actually determined by the average motion of all 'celestial bodies': a gyroscope's axis, for instance, would be fixed relative to the distant galaxies. What would happen to a rotating bucket, he speculated, if the rest of the universe were removed? It seemed to him that an inertial frame made no sense in an empty universe. Mach's principle, as it came to be called, has figured prominently in cosmological debates ever since.

Among the universes that satisfy the equations of general relativity, there are some in which the distant galaxies would very slowly move across the sky, relative to the axis of a gyroscope. If Mach were right, some extra 'principle' would restrict possible universes more stringently than Einstein's equations do, by ruling out the rotating solutions. Einstein took Mach's principle seriously, and deemed only the finite and 'closed' universes to be truly Machian. I learned this point of view from Wheeler, who argued forcefully that our universe must be dense enough to ensure recollapse.

Another appealing argument for a finite universe was that an ensemble of separate universes would be easier to stomach if each member of the ensemble was finite rather than infinite.

COUNTDOWN TO A BIG CRUNCH?

The closed or recollapsing universe was certainly my favoured assumption back in 1969, when I wrote a short paper pretentiously entitled 'The collapse of the universe: an eschatological study' about what will happen if our universe collapses.

Suppose the present density were twice the critical value (or, in other words, omega were 2). The expansion would stop when the galaxies were twice as far away from each other as they are today. Thereafter, they would fall towards one another, their redshifts being replaced by blueshifts. Space is already punctured by black holes created when massive stars die, or by the runaway collapses in galactic centres that manifest themselves as quasars. But these would then just be precursors of a universal squeeze that eventually engulfs everything.

About 100 million years before the crunch, individual galaxies would merge. Later in the 'countdown' the remaining stars, no longer attached to their parent galaxy, would become dispersed throughout the contracting universe. They'd move faster as the contraction progressed, just as atoms in a box move faster (and the gas heats up) when the box is compressed. Eventually, the stars would be shattered by colliding with each other. I was surprised, though, to calculate that most stars would have been destroyed in another way before they got close-packed enough to collide. The sky, brightened by the blueshifted radiation from all other stars (plus the primordial background radiation, which would itself heat up during the compression) would become hotter than the stars themselves. Stars would get cooked in an over even hotter than their surfaces, and soak up heat faster than they could get rid of it; they would 'puff up' and disperse into gas.

The earliest this could happen would be 50 billion years from now; the breathing space is at least 10 times the remaining lifetime of the Sun. The final outcome would be a fireball resembling that with which our expansion began.

But the collapse would not be *exactly* like the initial big bang with the direction of time reversed. The early universe was smooth and uniform, except for the small 'ripples' that evolved into galaxies and clusters; in contrast, the 'crunch' would be irregular and unsynchronised. Our universe is developing more and more structure as it expands, and this trend would not reverse during the contraction phase. Anything that has fallen into a black hole has already, in effect, experienced the final crunch; even more black holes will form during the contraction, and material will experience violent shearing motions. Roger Penrose believes that this difference between the smooth initial stages and the irregular final stages is crucial in setting the arrow of time. (This is discussed further in Chapter 13.)

Could a collapsing universe rebound phoenix-like into a new cycle? Nothing could stop the density rising to infinity – to a 'singularity'. Such a singularity was once thought to be an artefact of the special symmetry and uniformity. If, for example, the stars in a cluster were pulled inward by gravity in a uniform and exactly symmetrical fashion, they would clearly all collide in the middle. However, small sideways motions would, in Newton's theory, prevent them from all converging to the same point; they may then miss each other, so that the cluster can re-expand to its original size. Perhaps something analogous could happen to a collapsing universe. But Hawking and Penrose showed that Newtonian intuitions are misleading. Singularities are inevitable even when the collapse is irregular: the kinetic energy itself (since energy is equivalent to mass) exerts extra gravity, so the attraction feeds on itself.

Physical conditions in the 'bounce' would transcend the physics we understand, so that nothing could be said about the possibility of a rebound into a new cycle – still less about what memory would be preserved of what had gone before. The concept of an 'arrow of time' – what is 'before' and what is 'after' – breaks down under these extreme conditions.

PERPETUAL EXPANSION

What about the other case, when there isn't enough gravitating stuff to halt the expansion? Our universe would then have time to run down to a final heat death. If a cosmologist had to answer the question, 'What is happening in our universe?', in just one sentence, a good answer would be: 'Gravitational energy is being released, as stars, galaxies and clusters progressively contract, this inexorable trend being delayed by rotation, nuclear energy and the sheer scale of astronomical systems, which makes things happen slowly and staves off gravity's final victory.' If our universe expands for ever, there *will* be enough time for all stars, all galaxies, to attain a terminal equilibrium.

The sky would get still blacker as galaxies dispersed ever more thinly through the expanding space. But our universe would darken for another reason as well. Galaxies would get intrinsically dimmer. Their constituent atoms would continue to be recycled through successive generations of stars: hydrogen would be processed into helium (and further up the periodic table). Bright stars cannot be created from already-spent fuel. More and more gas would get locked up, either in faint stars of very low mass, or in dead remnants: neutron stars, white dwarfs or black holes.

Just as our Milky Way will crash into Andromeda, so most galaxies will combine with others in the same group or cluster. Each supercluster will become one unit: the black holes at the centre of each galaxy will sink into the middle of the merged system, surrounded by a swarm of dead stars. The hierarchical clustering that has already led to galaxies, clusters and superclusters will continue to still grander scales.[2]

ATOMS ARE NOT FOR EVER

Most of the atoms that went into the making of galaxies will eventually get trapped in black holes or inert stellar remnants; each galaxy will become just a dark swarm of cooled white dwarfs, neutron stars and black holes. But eventually the atoms will themselves decay: if the baryons of which they are made were absolutely immutable (as we believe the amount of electric charge in our universe is), the excess of matter over antimatter could never have emerged in the ultra-early universe. The eventual decay of protons restores the symmetry between matter and antimatter with which our universe began.

The average lifetime of an atom exceeds the present Hubble time (the present 'age' of our universe) by more than 20 powers of 10: even though one tonne of material contains about 10^{30} atoms, experimenters would need to watch many tonnes (maybe even many thousands of tonnes) to detect even one decay in the course of a year. But, given enough time, white dwarfs and neutron stars would dissolve away; so would any diffuse intergalactic gas. The energy would go into electrons and neutrinos.

Black holes would be unaffected by proton decay. But even they do not live for ever. They shed energy by the 'quantum evaporation' process (see Chapter 11). This process could be important – and, indeed, conspicuous – in the present-day universe if 'miniholes' were created by the huge pressures of the ultra-early universe. Tiny black holes, weighing as much as a mountain but only the size of an atomic nucleus, radiate intensely; as they lose energy (and mass) they shrink but radiate even more energetically, and may eventually disappear in a final outburst of particles and gamma rays.

Bigger black holes are cooler and radiate more slowly. The holes that form when heavy stars die would take 10^{66} years to evaporate. But an ever-expanding universe provides enough time for this to happen – enough time, even, for evaporation of the ultra-massive black holes, each as heavy as millions of stars, that build up in the centres of galaxies or supergalaxies (and which decay much more slowly still). Everything the holes had ever swallowed would thereby be recycled back into radiation.

If even the heaviest black holes eventually evaporated too, nothing would be left but radiation, and electrons and positrons. An electron could annihilate by colliding with a positron. A direct hit is highly improbable; electrons and positrons could nevertheless be brought together by forming a bound pair, orbiting around each other, and then spiralling together. So immensely dilute does everything become that there would eventually, on average, be less than one electron in a volume as large as our present observable universe. Immensely wide binary pairs could form: an electron's motion could be controlled by the electric field of a single positron 10 billion light-years away, and after enough aeons had passed the radiation drag would have brought them closer together.

CAN 'LIFE' SURVIVE FOR EVER?

The first thorough discussion of what might happen in an ever-expanding universe came in an article entitled 'Time without End:

Physics and Biology in an Open Universe'. This article was detailed and scholarly (in contrast to my own earlier one on the collapsing universe), and appeared in the austerely technical journal *Reviews of Modern Physics* (the adjacent article in the same issue has the more forbidding title 'Classical solutions of SU(2) Yang-Mills theory'). Its author, Freeman Dyson, combines, to a unique degree among physicists, formal mathematical brilliance with an enthusiasm for wide-ranging speculation. Dyson's scientific eminence dates from 1947, when he was still a student. The theory known as quantum electrodynamics – the most precise and successful theory in the whole of physics – was being developed by Richard Feynman and Julian Schwinger, using quite different approaches (as well as, quite independently, by Sin-itiro Tomonaga in Japan). Dyson showed how the very different mathematical ideas underlying Feynman and Schwinger's approaches could be linked together.

Dyson has spent most of his career as a professor at the Institute for Advanced Study at Princeton. At this unusual institution there are no students: the staff, free of the duties and constraints that encumber anyone working in a normal university or laboratory, are generously supported to pursue any research that takes their fancy. Although he has continued to study 'formal' mathematical aspects of physics, Dyson has latterly been more influential through his speculative ideas bordering on science fiction, and his eloquent books and lectures celebrating the diversity and complexity of our world. His career exemplifies the best justification (and it's hard to think of many) for a studentless intellectual haven like the Institute for Advanced Study.

Will our universe expand for ever? Dyson cannot tell us – indeed, we still cannot decide this issue – but he has no doubt which option he wants:

> The end of a closed universe has been studied in detail by Rees. Regrettably I have to concur that in this case we have no escape from frying. No matter how deep we burrow into the Earth to shield ourselves from blue-shifted background radiation, we can only postpone by a few million years our miserable end. ... It gives me a feeling of claustrophobia to imagine our whole existence confined within a box.

Dyson was writing after Hawking's work on black-hole evaporation, but before the case for proton decay became widely accepted. He therefore considered what might happen if, even after black holes have decayed, there were still white dwarfs and neutron stars. The final heat death would be spun out over a much longer period. But still not for ever. A neutron star could form a black hole by quantum tunnelling. This immensely improbable event – 10^{57} atoms 'quantum-jumping' in unison – would not be expected until after a time so enormous that, written out in full, it would require as many zeros as there are atoms in the observable universe! The resultant black holes would then evaporate, in a time that, in this perspective, is almost instantaneous.

But what is the prognosis for some exotic form of intelligent life? Even after the stars have died (precluding any manifestations of life such as might evolve on Earth), can 'life' survive and develop intellectually for ever, thinking infinite thoughts and storing or communicating an ever-increasing body of information? 'Energy reserves' are finite. However, it takes smaller pulses of energy to store or transmit information if this is done at a low temperature. And the universe will cool down as it expands. Instead of being 2.7 degrees above absolute zero, it will, after a trillion years, be below a millidegree. As the background temperature falls, any conceivable form of life or intelligence would have to keep cool, think progressively more slowly and hibernate for long intervals.

If protons lasted for ever, hugely complex but tenuous networks could be fabricated. There are quite general limitations on the size and complexity of organisms (or, indeed, computers) because anything too heavy would be crushed by gravity, and its internal workings would generate too much power to be radiated away. But structures in the far future can transcend both these constraints. Gravity can be suppressed, however massive these constructions are, by making them distended enough. And they can have a large enough surface to radiate and stay almost as cool as the background radiation, whose temperature drops as the expansion proceeds; the minimum energy needed to transmit each item of information gets ever lower. Information-processing (or 'thinking') would be very slow in a spread-out configuration: the rate is limited by how long a signal takes to cross it, moving at the speed of light. (This is of course why supercomputers are made as compact as possible.) But what is the urgency when aeons stretch ahead?

Evolution needn't come to an end, even when all the protons

vanish. There could always be black holes, provided that they grew, by coalescence, fast enough to counteract their erosion by evaporation. (Their masses would need to go up as fast as the cube root of time – growing tenfold for each factor 1000 on the cosmic clock.) These holes may concentrate energy enough to create new matter. Even a dilute gas made of electrons and positrons could provide the basis for circuitry controlled by complex magnetic fields and currents pervading the medium. This is reminiscent of the inorganic intelligence depicted in *The Black Cloud*, the first of Fred Hoyle's science-fiction novels (and the most carefully crafted and imaginative of them).

SUBJECTIVE ETERNITY BEFORE THE CRUNCH?

Even if 'physical' time runs on for ever, Dyson's reassuring conclusion that an infinite amount of *subjective* time lies ahead shouldn't be dismissed as obvious. The 'endgame' of a perpetually expanding universe is played out more and more slowly: any elemental action – a 'thought', or the processing of a single bit of information – takes longer and longer. Just as an infinite series of numbers (for instance, 1, ½, ¼, ⅛ . . .) can add up to a finite sum, there might have turned out to be a finite bound on 'subjective' time.

The 'time' that appears as the symbol *t* in the equations of physics is not necessarily a good measure of the 'time' whose passage is marked by a succession of significant events.[3] In a perpetually-expanding universe, the pace of activity gets ever slower. Perhaps, conversely, this distinction between 'subjective' and physical time allows us to view a *big crunch* more optimistically.

Time is measured by the ticking of standardised clocks. However, no conceivable clock could survive the final stages of a big crunch; and any clock falling into a black hole would be shredded by tidal forces before encountering the 'singularity'. We might start off measuring time in years, by orbits of planets around stars. But, in the later stages of the 'countdown' to the crunch, each star's surroundings become more crowded: stars would be hurtling so close that no planets could survive undisturbed in their orbits, and we should have to use atomic clocks. And the atoms themselves would be destroyed eventually. As conditions became more extreme, timekeeping would require successively smaller and sturdier clocks. No finite sequence of clocks could record every instant right up to the 'singularity'.

This recalls Zeno's classic paradox on the 'impossibility' of motion: before completing a journey, you must get halfway: before that, you must get a quarter of the way . . . and so on; there are an infinite number of things to do in getting started. Though seemingly paradoxical, this claim about a collapsing universe isn't manifestly as fallacious as Zeno's argument. Unlike in our everyday world, there is no natural clock that can be used all the time. Instead, an infinite series of increasingly fast-ticking clocks would need to be used in succession.

Viewed in this perspective, the final singularity seems a remote abstraction, separated from us by an infinite number of intervening events. John Barrow and Frank Tipler have developed this line of thought to its limits. If the crunch were smooth and symmetrical, like the time-reverse of the big bang, there would be no possibility of infinite subjective time. This follows as a corollary of Dyson's line of argument for an ever-expanding universe: Dyson found that finite energy reserves were no constraint, because energy could be used in smaller and smaller quanta as the universe cooled down; in contrast, the required quanta get ever larger (and energy is used less and less efficiently) as the universe compresses and heats up.

Barrow and Tipler claim that the prospects for infinite 'subjective time' are better if the collapse occurs in a 'skew' or anisotropic fashion. Their argument builds on ideas developed in the late 1960s by Charles Misner, who showed that anisotropic universes display what he termed 'mixmaster' behaviour: strong shearing motions, such that a contracting universe would be squeezed alternately in different directions. This shearing motion can generate enough energy to supply arbitrarily many quanta, despite the requirement that these quanta must get larger as collapse proceeds. (The intervention of quantum effects on the gravitational field itself, however, may 'choke off' this process and preclude any infinite regress.) Tipler speculates that an anisotropic 'crunch' offers a propitious environment for complex structures with at least some attributes of life, provided that our universe continues for 10^{15} years before recollapsing, so as to allow enough time to prepare for it!

'FAST FORWARD' INTO THE FUTURE

Will our descendants need to follow Dyson's conservationist maxims to survive an infinite future? Or, at the opposite extreme, will they fry in the big crunch a few tens of billion years hence? We shall need to compile a more complete inventory of what is in the universe, by observing in all wavebands, and searching for all forms of possible 'dark matter', before we can pronounce a more reliable long-range forecast for the next 100 billion years and beyond.

If you're of an apocalyptic temperament and can't wait 100 billion years, then head for a black hole – you'll there encounter a 'singularity', created by a local gravitational collapse prefiguring the big crunch. Black holes form when heavy stars die – perhaps after some supernovae – and there are many millions of them within our Milky Way. Monster black holes each weighing as much as a billion suns lurk in the centres of some galaxies – these are the relic of a catastrophic quasar phase when the galaxy was young. You should aim, preferably, for one of these monster holes: they are so capacious that, even after passing inside, several free-falling hours would remain for leisured observation before the extreme gravitational stresses near the central 'singularity' shredded you apart. If the black hole is spinning, careful navigation may evade the singularity.

A more prudent course would be to remain just outside the hole. The closest orbits around a rapidly spinning black hole have the remarkable property that the time-dilation can be arbitrarily large. A clock moving in such an orbit would seem, to a distant observer, to be deeply redshifted, and running exceedingly slow. Conversely, someone in such an orbit would get a speeded-up preview of the entire future of the external universe.

TERRESTRIAL AND COSMIC HAZARDS

Our biosphere has taken 4.5 billion years to evolve. But the Solar System has another 5 billion years ahead of it. Our universe, even if it eventually recollapses, still has at least 90 per cent of its course to run. And, if it expands for ever, there may be infinite time and infinite space in which life can develop. In this cosmic perspective, we are still near the beginning of the evolutionary process. The progression towards complexity and diversity has much further to go. There is time enough for life to spread from our Earth through the entire Galaxy, and even beyond. And it may be, primarily, collective human actions that will determine how, or even if, that process unfolds. If Earth's biosphere were snuffed out, potentialities of truly cosmic proportions would be foreclosed. Being mindful of these potentialities stretches our horizons: it may even deepen our commitment to understanding our world, and conserving its web of life.

The Earth has always been vulnerable to catastrophes. Many species may well have succumbed to plagues. Impacts of asteroids and comets have caused worldwide extinctions. There is more than one chance in 100 000 that, within the next 50 years, the Earth will be hit by an asteroid large enough to cause worldwide devastation – ocean waves hundreds of feet high, tremendous earthquakes and changes in global weather. The mortality risk is low, but no lower (for the average person) than the risk, per year, of being killed in an air crash. Indeed, it's higher than any other natural hazards that most Europeans or North Americans are exposed to.

Added to these is now the risk of manmade catastrophe. The nuclear peril may loom less large than it did during the Cold War but such dangers could resurface. And there could be biological dangers – in this case not organised warfare, but perpetrated by terrorism, or even accident. We are used to the idea that a software 'virus' can spread through a computer network. Perhaps an artificially engineered virus could cause a worldwide epidemic. The risk of all human life being extinguished may be slight. But, with such technology readily accessible to terrorist groups (or, indeed, to innocent experimenters who may unleash catastrophe quite unwittingly), could one confidently assess the risk, during the next century, as less than 10 per cent? And what about the century after that?

The *scientific* case for manned spaceflight gets ever weaker as robotic and miniaturisation techniques develop. Experiments and planetary exploration are better (and far more cheaply) carried out by fleets of tiny unmanned probes. Even in the heyday of the Apollo programme, the scientific justification was slender. We learned as much about the early Solar System from meteorites that crashed into the Earth as from analysing moondust. Apollo was driven by superpower rivalry, and the 'spectator-sport' dimension. The *last* lunar landing was in 1972. To a whole new generation, 'men on the moon' are scenes in a remote historical episode, driven by motives almost as strange as those that led to the pyramids. The Apollo programme was seen as an end in itself, rather than a step towards some inspiring longer-range goal that might have sustained its momentum.

The strongest such case – indeed perhaps the only credible one – is that manned spaceflight offers a global (or even cosmic) insurance policy. The ever-present risk from nature has been augmented since humankind entered the nuclear and biotechnological age. The species will remain vulnerable to these (probably increasing) hazards so long as it is confined here on Earth. Self-sustaining communities away from the Earth could be developed within a century if current momentum in space technology were maintained. But is it realistic to divert resources from more immediate needs to guard against a risk (one per cent? ten per cent?) that humans will become extinct, thereby eliminating a potential that stretches ahead not just for a historical timespan, but for as long as we've taken to evolve from protozoa – perhaps, indeed, far longer still. Ecological issues – the fragility of the Earth's biological diversity, and the importance of preserving it – now loom large in public awareness. Perhaps these concerns with the potential and destiny of terrestrial life will engender renewed commitment to manned spaceflights.

A PREMATURE APOCALYPSE?

We need not fret about the collapse of our entire universe: this will happen, if at all, long after the Sun has died. But what if humans could pre-empt this ultimate 'natural' catastrophe, and inadvertently unleash a disaster that destroyed not just life on Earth, but the entire cosmos? This scenario is a conceivable (though fortunately unlikely) consequence of current concepts about the fundamental forces.

As our universe cools down, space itself – what physicists call 'the vacuum' – changes its nature, just as steam does when it condenses into liquid water, and then into ice. The 'vacuum' is the arena in which all the particles and forces interact; when it undergoes a 'phase transition', the masses of particles and the forces between them alter drastically. According to the generally accepted theory of Salam and Weinberg, such a transition occurred when our universe was about 10^{-12} seconds old. Before that time, the forces of electromagnetism and the so-called 'weak' force (which operates in radioactivity and in reactions involving neutrinos) were combined in a single 'electroweak' force; it took a phase transition to give them their present properties. Attempts to extend this unification to embrace the strong (or nuclear) forces as well are still tentative. But these 'grand unified theories' predict another still earlier phase transition, when the universe was only 10^{-36} seconds old, before which the forces governing the microworld (that is to say, all the fundamental interactions except for gravity) would be combined in a single primeval force.

If there have already been two phase transitions, in each of which the vacuum 'cooled' to a lower-energy state, could there be more? The Harvard theorist Sidney Coleman conjectured that there could be a third transition, *which may not yet have happened*. The present vacuum could be supercooled, rather as very pure water can remain liquid below freezing point; a change in the vacuum could then be triggered by some local concentration of energy, just as a speck of dust can induce supercooled water to crystallise suddenly into ice. Could physical laws governing the entire universe be transmogrified, simply by the 'trigger' of a local energetic event?

Creating the greatest possible concentration of energy is the main purpose of the huge machines used by physicists at CERN, Fermilab and elsewhere: particles are accelerated to high energies, and then crashed into one another. Is there a risk that the next generation of machines could inadvertently tear the fabric of space? This would be the ultimate catastrophe. A bubble of the new vacuum would expand at the speed of light. We would have no advance warning of its approach; we wouldn't know what had hit us. The bubble wall would surge onward until it engulfed our entire universe. Worse still, the 'new' universe inside the bubble would be a 'stillborn' one in which nothing could evolve: it would behave like the time-reverse of an inflationary universe, squeezing everything towards infinite density.

Before the first nuclear tests were carried out, physicists prudently calculated, with reassuring results, that the deuterium in the world's oceans wouldn't be ignited by a thermonuclear explosion. What Sidney Coleman envisaged would be a cosmic and not merely terrestrial catastrophe.

At Princeton's Institute for Advanced Study, the aura of Freeman Dyson encourages unrestrained speculation. During a visit there, I discussed the risk of such a cosmic catastrophe with Piet Hut, one of Dyson's younger colleagues. Hut had previously studied a non-nuclear kind of terrestrial catastrophe. With some colleagues at Berkeley, he had conjectured that a faint star was trapped in an exceedingly large eccentric orbit around the Sun; every 35 million years or so, the orbit would plunge into the Solar System, perturbing comets and triggering a shower of impacts that wreaked havoc on Earth. (The extinction of the dinosaurs was blamed on such an impact.) There is incontrovertible evidence for (not necessarily periodic) impacts of comets or asteroids: one of these may indeed have wiped out the dinosaurs; in future, we could suffer a similar fate. But this so-called Nemesis theory has now been abandoned – the proposed orbit was interesting, but no star seemed to occupy it.

Could unwary experimenters crash particles together with enough energy to transform the vacuum into a new state, triggering an expanding 'bubble' that would destroy our universe? Piet Hut and I did some simple calculations. We wanted to check whether particle accelerators in the laboratory could generate more energetic impacts than have happened naturally. The most energetic natural collisions involve 'cosmic rays' – very fast particles that pervade the entire Galaxy, perhaps even the entire observable universe. These particles have been accelerated, almost to the speed of light, by cosmic explosions even more violent than supernovae explosions: they could come from the strong radio sources discussed in Chapter 2, which we now think are energised by jets from huge black holes. Or they could come from the sudden release of energy when (for instance) two neutron stars orbiting each other spiral together and coalesce. Their origin is a mystery, but we know from direct measurements that the fastest cosmic ray particles reaching us from space are millions of times more energetic than any that can be artificially accelerated here on Earth. Each of them, just a single atom, carries as much punch as a bullet or a fast-served tennis ball.

Cosmic rays with these extreme energies are so exceedingly rare – an area of one square kilometre would intercept only one per century – that no two would ever have actually collided with each other anywhere in our universe. However, we calculated that there would have been many collisions between cosmic rays whose energies, though somewhat less extreme, were still 100 times higher than can be attained by the Large Hadron Collider, the world's most powerful accelerator now being built at CERN in Geneva. So any forseeable laboratory-induced collisions would be relatively gentle compared with those that have occurred repeatedly, without catastrophic consequences, throughout interstellar space.

This wasn't an entirely frivolous calculation. The vulnerability of space to a catastrophic phase transition was admittedly just the figment of a speculative theory. But the possibility is not absurd – in our present state of ignorance about unified theories, we would be imprudent to disregard it. Indeed, caution should surely be urged (if not enforced) on experiments that created energy concentrations that may never have occurred naturally. We can only hope that extraterrestrials with greater technical resources, should they exist, are equally cautious!

NOTES

1 In our Galaxy, most stars are in a thin disc, and each follows a near-circular orbit around the Galactic Centre; the stars are held in these orbits by the pull of all the other stars (and the dark matter) which collectively balance centrifugal force. When the merger occurs, each star in our own Galaxy will feel just as strong a force from Andromeda. But this latter force, acting obliquely to our own disc, will therefore disorganise the stellar orbits in our Galaxy. Conversely, the Milky Way's gravity will deflect all Andromeda's stars out of its disc.

2 Our present universe is slightly irregular on scales even larger than superclusters. Just as the precursors of galaxies and clusters were small-amplitude ripples in the early universe, so aggregates even larger than superclusters may eventually condense from regions that are now only slightly denser than average. A region one per cent denser than the average will condense out when the universe has expanded by a further factor of 100; if the overdensity is only 0.1 per cent, the universe must expand by a factor 1000, and so on. We know about these overdensities on very large scales because they cause slight non-uniformities in the microwave background temperature revealed by the COBE satellite. The hierarchy of clustering will grow in step with the age of the universe. The largest superclusters have dimensions that are one per cent of the Hubble radius (they contain maybe a millionth of the total mass within our 'horizon'). When the 'horizon' is bigger, so also will be the largest clusters. If omega is exactly 1, and the ripples are the same on every scale (the 'Harrison–Zeldovich' assumption described in Chapter 7), the biggest clusters will grow in step with the Hubble radius. If the 'ripples' had larger amplitudes on larger scales, the universe may eventually close up on itself, because the

scale of the largest structures becomes as large as the horizon. On the other hand, the ripples may get weaker on larger scales, in which case structures could at some stage stop growing and merging. The most drastic difference would occur if there were a cosmological constant lambda (as discussed in Chapter 8). Even if this were too small to affect the cosmic expansion now, it would eventually (unless it were exactly zero) exert a cosmic repulsion that overwhelmed the ever-diluting strength of gravity. An observer on any cluster would see the others disperse with increasing speed, until none at all were left within view.

3 An anonymous graffiti-writer at the University of Texas is credited with this insight: 'Time is nature's way of stopping things from happening all at once.'

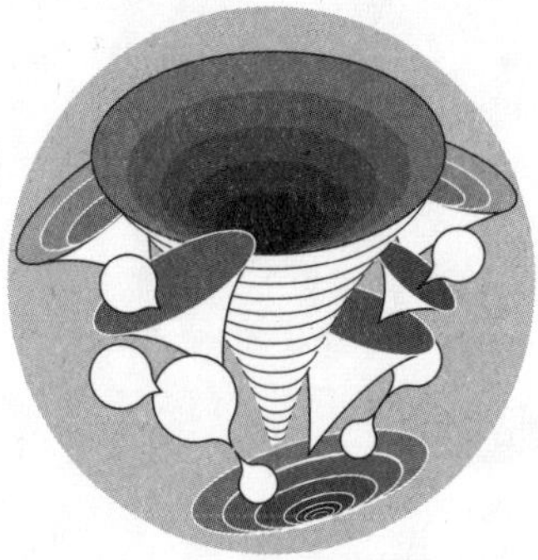

13

TIME IN OTHER UNIVERSES

The clock's ticking away here at Wembley – it's ticking away everywhere, I guess.

Football commentator

CLOCKS TICK AT DIFFERENT RATES, depending on where they are or how they move. Although surprising, such a conclusion is forced on us, as Einstein realised, by an amazing experimental fact – that if you measure the speed of light the answer is the same whether you move towards or away from its source. We experience three spatial dimensions: left and right; forwards and backwards; up and down. Three numbers are therefore needed to locate an event in space. We also need a fourth number to define an event: the *time*, measured on some kind of clock. Time is clearly different from the other three dimensions: we have some freedom to move in the three spatial dimensions, but are carried along willy-nilly by the 'flow' of time. Einstein showed that time and space are linked. Two events don't have a uniquely defined separation in space, nor a uniquely defined time interval between them. Instead, they have a definite separation in four-dimensional space-time; their distance apart in space, and their separation in time, depend on how the observer is moving.

Can time be cyclic? Is there a universal 'arrow' that distinguishes the past from the future? Does nature prohibit something so paradoxical as 'time travel' into the past? And are there limits to how long 'time' can last, or how finely it can be measured?

Time may not 'tick away' for ever. The big bang could have been the 'beginning' of time and space, as well as the origin of all the material filling our universe. We cannot extrapolate 'time's arrow' back before the big bang, nor (if our universe recollapses) after the big crunch. Whatever happens to our universe, anyone who fell into a black hole would experience a finite future (indeed, the 'curtain' could fall disconcertingly fast).

The *longest* timespan that is meaningful – the interval between big bang and final crunch – is, at the very least, several tens of billions of years. But what about the converse question: is there a *smallest* scale? Can time be 'chopped up' into indefinitely small intervals? Quantum physics supplies an answer. Heisenberg's uncertainty relationship tells us that, to measure a time interval with increasing precision, we need to use quanta of shorter wavelength (and thus higher energy). Because the light quanta move at finite speed, this increasing amount of energy must be focused into smaller dimensions (smaller than the interval being measured multiplied by the speed of light). A limit arises when the requisite energy is so high, and so tightly concentrated, that it would collapse to a black hole. Quantifying this argument suggests that there is a minimum timescale of about 10^{-43} seconds, commonly called the Planck time. Events cannot be timed or ordered more precisely than this.[1]

Near the beginning (or end) of the universe, everything would be squeezed into an exotic state that mixes up the dimensions of 'space' and 'time'. Even if we were confident enough about the physics of the ultra-early universe (see Chapter 9) to extrapolate back close to the Planck time, there is then a firm barrier. On this tiny scale, some theories, going back to Wheeler's pioneering ideas in the 1950s, suggest that the time dimension was intermingled with the three spatial dimensions into a froth of 'space-time foam'. According to currently popular superstring theories, there may be six extra dimensions. Space is tightly 'rolled up' in these extra dimensions, so they manifest themselves only on very tiny scales.

Hartle and Hawking have developed a different approach to 'quantum cosmology' and the beginning of time. They suggest that the distinction between time and space was initially blurred, so that 'What happens before the big bang?' becomes a question akin to 'What happens if you journey north from the North Pole?'

Our concepts of space and time derive from experience and perceptions in the everyday world. We shouldn't be surprised that our intuition fails on either the cosmic or the submicroscopic scale – what is remarkable (and a gratifying simplification) is that 'commonsense' concepts remain valid over as large a range as they do.

Most of us find an infinite past (as, for instance, there would have been in a steady-state universe) harder to envisage than an infinite future – we feel there must have been a beginning, but not necessarily an end. This isn't just a matter of psychology: physics offers very good reasons for reacting differently to these two prospects. The famous second law of thermodynamics tells us that systems become more disordered as time goes on, that hot and cold bodies gradually equilibrate, and so forth. If an infinite past has already elapsed, why hasn't everything 'run down' already? This argument would be a cogent one when applied to a static bounded system in a box, but carries less conviction in an open, possibly infinite, and dynamic system like our universe.

If we can't confidently extend to an entire universe our everyday experience (where thermodynamics, our own memories and many other phenomena manifestly distinguish a direction of time), we confront a new fundamental question: what distinguishes 'past' and 'future' on the cosmic scale? Does a universal 'arrow of time' point unambiguously towards the future?

TIME'S ARROW

With one trivial-seeming exception (the K-meson decays mentioned in Chapter 9) the laws governing the *micro*world are reversible in time: a movie showing atomic particles colliding with each other would generally look no different when run backwards. The *macro*scopic physical world, in contrast, manifests a definite arrow of time, set by increasing disorder (or entropy). We are used to seeing waves moving out from a source – for example, when a stone is thrown into a pond – but would be amazed by waterwaves that converged inward to a focus. Likewise for the laws of electricity and magnetism: an electric charge radiates when it is shaken; artificial radio waves are generated by streams of moving electrons in a transmitter. But electromagnetic waves never focus on to a charge previously at rest and set it oscillating. Our subjective perception of time is obviously asymmetric. We have memories of the past only, and hindcasting is in general more reliable than forecasting (but not always: it is easier, for instance, to predict when artificial satellites in low orbit will burn up in the atmosphere, than to infer when or where they were launched).

Throughout our ordinary experience, time's arrow is utterly unambiguous. Any movie of everyday (*macro*scopic) events looks grotesquely different when run backwards. Cause and effect are reversed – causal explanations are replaced by 'teleological' interpretations. Broken bits of glass, and drops of liquid, seem to rush purposefully together in order to assemble into a glassful of wine in your hand.

The various local phenomena that distinguish past and future – the increase of entropy, the asymmetry between the way radiation is emitted and absorbed, and the fact that we remember only the past – are connected to each other. But cosmologists recognise another way of defining time's arrow: in terms of the expansion of the universe. This singles out a particular direction in time; it also imposes a finite timespan (back into the past if the expansion continues indefinitely; in both past and future directions if our universe recollapses). Is this cosmic arrow of time irrelevant in the everyday world, or could the difference between past and future actually be *imposed* by the way our entire universe behaves?

ENTROPY AND IRREVERSIBILITY

The everyday world is very far from thermal equilibrium – there are enormous contrasts between hot and cold. It is not completely ordered; nor has it 'run down' to a completely disordered and random state. The same is true for the cosmos on larger scales – there are huge contrasts between the stars with their blazing surfaces (and still hotter centres) and the sky between them, which is almost at the 'absolute zero' of temperature – not quite, of course, because it is warmed to 2.7 degrees by the microwave 'echos' from the big bang. In the ultimate future, as discussed in Chapter 12, conditions may revert closer to equilibrium, but this will take immensely long even compared with the universe's present age.

Our universe started off hot and almost featureless. That it should have evolved *away* from equilibrium towards a manifestly more structured state is at first sight perplexing – indeed it may seem to conflict with the basic thermodynamic idea that everything 'runs down'. What accounts for the emergence of complex cosmic structure?

The two prerequisites are (i) *expansion*, which sets a well-defined asymmetry between past and future, and (ii) *gravity*, which allows density contrasts to grow, leading to eventual emergence of structures, as the universe expands.

If all the microscopic processes – collisions between particles, emission and absorption of photons and so on – were very fast compared with the expansion rate, everything would, at every instant, be in equilibrium. The material would retain no 'memory' of whether it had previously been denser or less dense, and would bear no imprint of the direction of time. But, as our universe expands and becomes more dilute, these reactions get slower and eventually become ineffectual.

For example, if nuclear reactions happened faster (or the expansion had been slower) all the primordial material would have been 'cooked' into iron, as inside a very hot star. No stars could then have existed in our present universe, because all the available nuclear energy would already have been used up in the early fireball. Fortunately (as described in Chapter 3) the first few minutes of the expansion didn't allow enough time for the reactions to do more than convert about 25 per cent of the hydrogen into helium – the outcome would be quite different if the universe were contracting. This exemplifies the crucial importance of cosmic expansion.

Another irreversible effect, as Sakharov first pointed out, established an excess of matter over antimatter at a still earlier stage (see Chapter 9). Had that not occurred, all the matter would have annihilated with an equal amount of antimatter, leaving a universe containing no atoms at all – there would then be no stars, still less any of the chemistry that has allowed complex structures to emerge.

GRAVITY'S THERMAL PERVERSITY

Gravity is even more crucial. Suppose a universe were expanding at the same rate as ours, but gravity had been switched off. Then there would be nothing to perturb the uniformity of the material, which would now, after 10–20 billion years, be a cold dilute gas uniformly pervading all of space. It is gravity that renders uniform universes unstable, and allows large density contrasts to develop from small initial irregularities (see Chapter 7). Vast protogalactic gas clouds can thereby separate out and then fragment into stars.

Gravity is a dominant influence on the stars themselves. Stars have the property, at first sight an odd one, that they *heat up* when they *lose* energy. Suppose that the fuel in the Sun's centre were switched off. Heat would continue to leak out, and be radiated away from its surface. If nuclear fusion didn't regenerate this energy, the Sun would contract. But it would then end up with a *hotter* centre than before: to establish a new and more compact equilibrium where the central pressure is high enough to balance a (now stronger) gravitational force, the internal temperature must rise. Anyone who learned physics at school (especially if the style of teaching was old-fashioned) will have measured the 'specific heat' of a lump of metal, by dropping it into hot water and noting how the thermometer drops as the metal 'soaks up' heat from the water. Stars (and, indeed, any objects held together by gravity) have a specific heat that is *negative* – they need to lose (rather than gain) energy to get hotter.

Gravity has similarly counterintuitive effects when an artificial satellite in a low orbit feels atmospheric drag. As it spirals in towards burn-up, its orbital speed actually *rises* (a satellite in lower orbit feels the Earth's gravity more strongly and has to move faster). Only half the energy it loses is dissipated into heat; the other half goes into speeding it up.

Once systems form that are heavy enough to be self-gravitating, departures from equilibrium *grow*. Our universe can thus have evolved, quite consistently with thermodynamics, from a primordial fireball – hot, but *uniformly* hot – into a structured state containing very hot stars radiating into very cold empty space. As the universal expansion continues, disparities in density get more and more conspicuous. Individual stars become denser as they evolve (some ending as neutron stars or black holes), whereas overall the matter gets more thinly spread.

We're familiar with the idea that 'order' emerges in the Earth's biosphere because sunlight energises the elaborate chemistry (photosynthesis) that allows plants to grow. The Sun emits 'high-grade energy'. This is processed on the cooler Earth, and the waste heat then escapes into (still cooler) space. These temperature contrasts – prerequisites for the emergence of complexity – are the natural outcome of a chain of events that cosmologists can trace back to an ultradense primal medium that was almost structureless.

These beginnings are still, of course, mysterious, but there is no mystery about how the large-scale dominance of gravity could, at least in principle, lead to the differentiated structures that we now see, and which set the stage for intricate cosmic evolution, and the emergence of 'self-organising systems' such as ourselves. Just two things are required. First, very slight irregularities in the early universe (which may be no more than the tiny vibrations implied by quantum physics) must be present, to initiate gravitational instability and clustering. Second, there must be systems capable of absorbing energy from hot regions, and processing and reradiating it at lower temperatures.

THE BIG BRUNCH

If our universe recollapses, thermal equilibrium can't be restored until everything recompresses to such a high density that it becomes opaque. Until that stage, there is no obvious reason why local arrows of time should be affected. If the recollapse were to happen before stars had all expired, our remote descendants could find themselves in a universe where cosmic and local arrows of time pointed different ways. On the other hand, if collapse were postponed for enormously more than 10 billion years, not only would all stars have died, but all atoms and black holes would have dissolved into radiation. This makes the odds much better than evens that *any* observers would find their universe expanding.

Could time's arrow be even more intimately linked to the expansion? The equations of electricity and magnetism look symmetrical between past and future – they don't in themselves tell us why radiation spreads outward rather than converging inward. Some cosmologists, following pioneering work by Richard Feynman and John Wheeler, have suggested that cosmic expansion induces this asymmetry – because, in technical jargon, the radiation then encounters different *boundary conditions* in the future from in the past. This wouldn't lead to anything paradoxical in an ever-expanding universe. But, if we were in a closed, finite universe, there would then be no overall arrow from 'bang' to 'crunch': instead, time's arrow might point towards the direction in which the universe became larger, reversing at the moment of turnaround.

If the direction of psychological time reversed in a contracting universe, any conscious beings would of course perceive an apparently expanding universe, just as we do. Hawking briefly revived this idea in the 1980s. His attempts to apply quantum theory to the entire universe led him to envisage a closed 'cycle', in which the initial and final states were on the same footing. At first sight, this seemed to involve a reversal of time's arrow at 'turnaround'; however his colleague Don Page pointed out what Hawking later admitted was an error – the time-symmetry applied to the ensemble of universes, but not to a particular member of the ensemble.

Fascinating though the prospect might be, there seems no merit in the conjecture that time's arrow reverses at the instant of 'turnaround'. If our universe halted in its expansion and started to recollapse, nothing drastic – indeed nothing readily discernible – would signal that moment. Nearby galaxies would start to display blueshifts, but distant galaxies would still look redshifted so long as we were still receiving light that set out from them long before the 'turnaround'.

Penrose thinks time's arrow is orientated by the difference between the dynamics of the big bang and the big crunch. Our universe emerged from the big bang in an amazingly homogeneous state. The crunch would be more chaotic and less synchronised[2] – indeed, regions that have already collapsed to black holes are, in effect, precursors of the big crunch. Penrose suggests that, in a closed universe where the 'singularities' at the two ends of time are different, time's arrow points from the simple singularity towards the one that is more complicated.

SLOWER AND FASTER TIME

Even though 'time's arrow' always seems to point in the same direction, time doesn't always pass at the same rate. The best-known example is Einstein's so-called 'twin paradox': the twin who goes on a long fast journey, and then returns, ages less than the one who stays at home. This is not really a 'paradox': the phenomenon is merely surprising because it requires conditions vastly remote from the everyday experiences that have moulded our common sense. All experimenters (using their own clocks) measure the same speed of light however they are moving. That fast-moving clocks run slow is a necessary consequence of this remarkable but well-attested fact.

This time dilation becomes arbitrarily large for speeds approaching that of light. One consequence, incidentally, is that there is no limit to how far you could travel in your lifetime, if you accelerate sufficiently close to the speed of light. However, if you go a billion light-years and afterwards return, then, however young you still feel, more than two billion years would have passed at home – relativity doesn't allow speeds faster than light relative to the stay-at-home clock.

Time is similarly 'stretched' where gravity is strong (as was mentioned in Chapter 5). To a distant observer, clocks on a neutron star would seem to be running 20–30 per cent slow. Clocks in orbit very close to a spinning black hole (or on special trajectories inside it) could display an arbitrarily large time-dilation; conversely an observer in such an orbit could view the external universe *speeded up* by a factor to which there is no firm limit.

Differences between the tick rates of clocks in different places, or which are moving differently, are of course barely discernible in the everyday world where gravity is weak and motions are exceedingly slow compared with the speed of light. Aircraft fly at about a millionth the speed of light. The predicted effect has shown up in precise clocks carried in aircraft or rockets. You could prolong your life by about a millisecond by spending the whole time flying around the world from west to east.[3]

But there would be nothing deeply paradoxical if such effects were larger. The short novel *Einstein's Dreams*, by Alan Lightman, himself an astrophysicist, fantasises about distortions of time in an everyday world:

> There is a place where time stands still ... As a traveller approaches this place from any direction, he moves more and more slowly. His heartbeats grow farther apart, his breathing slackens, his temperature drops, his thoughts diminish, until he reaches dead centre and stops. From this place, time travels outwards in concentric circles – at rest at the centre, slowly picking up speed at greater diameters.

TIME REVERSAL AND LOOPS IN TIME

Physicists speculate about hypothetical particles called tachyons, that travel faster than light. Such particles would be innocuous if they didn't interact with any ordinary material, but would then be of no interest since they couldn't (even in principle) transmit signals. On the other hand, if tachyons *could* send faster-than-light signals, they would lead to serious paradoxes. Einstein's theory tells us how to transform from one frame of reference to another. Viewed from some frames, a tachyonic signal would seem to arrive *before* it was sent. This raises the paradoxes of 'time machines'.

Any alteration in the *order* of events (rather than just the rate at which they happened) would be deeply paradoxical. If it were possible to traverse a 'time loop' and return to one's own past, obvious conundrums arise regarding causality and free will. Shooting dinosaurs may not lead to any contradiction, but to strangle one's grandmother in her cradle would lead to problems not merely of ethics but of causality. In Isaac Asimov's novel *The End of Eternity* such inconsistencies are stopped by the actions of the 'time police'. Lightman's time-traveller in *Einstein's Dreams* is a figure of pathos: 'If he makes the slightest alteration in anything, he may destroy the future ... he is an exile in time.'

Commonsense intuitions are based on our experience of 'ordinary' lengthscales and timespans. We're ready to accept that everything is quite different on the atomic scale – what could be less 'intuitive' than quantum mechanics? And on smaller scales still, as already mentioned, space and time may be churned up in ways we cannot yet envisage. So can we confidently rule out 'time travel' on tiny scales?

The 'spacewarp', familiar from science fiction, allows instantaneous transport to great distances. But faster-than-light travel leads to the same problems as signalling with tachyons: from some frames of reference the traveller would be seen 'arriving' before 'departure'. In other words, if you could make a spacewarp, you could make a time machine as well. The physical concept that most closely resembles a spacewarp is a so-called 'wormhole' – a black hole linked umbilically to a very remote 'white hole'. Kip Thorne, Igor Novikov and their collaborators have explored such ideas very seriously. A key issue is whether the ends of the wormhole would close off before anything could traverse it. They have proved that there is no chance of holding a wormhole open unless it is made of material with very exotic properties – a very high negative pressure (in other words, tension). This kind of stuff could have existed at the huge energies of the ultra-early universe.[4]

Nobody has rigorously proved that it is utterly impossible to create, in our present universe, the kind of warped space-time that would allow time travel. But even if not absolutely prohibited, time machines would plainly be technical contrivances far beyond the imagination of H.G. Wells, whose 1895-vintage time-traveller appeared as 'a ghostly, indistinct figure sitting on a whirling mass of black and brass'.

Could there be closed time-loops on some microscopic scale? Would 'time machines' be so much stranger than the other strange paradoxes of the quantum world? The persistent ingenuity of theorists has still failed to settle whether 'wormholes' are merely technically unfeasible (just as, for instance, a 'time-stretching' spaceship moving at 99.99 per cent of the speed of light is technically unfeasible), or whether some fundamental law would absolutely prohibit them even if the requisite high-tension stuff existed from which a wormhole could be made. These issues are certainly not settled, but I would tend to believe the latter, and that physics has a kind of internal integrity that prohibits travel into the past. What Hawking has termed a 'chronology protection conjecture' would not only 'make the world safe for historians' and protect us from tourists from the future; it would, if valid, preclude time machines (microscopic or macroscopic) *even in principle*.

Time-loops stretching right around a universe were first discussed almost 50 years ago. The great logician Kurt Gödel achieved early fame when he showed that, even in a formal system as simple as arithmetic, statements could be written down that could never be proved either true or false within that system. He became a close friend of Einstein in his later years, when both worked in Princeton at the Institute for Advanced Study. (Gödel was a far more wayward and eccentric figure than Einstein himself. When Gödel formally took up American citizenship the enterprise almost came to grief when he tried to point out inadequacies in the US constitution; and he eventually died from malnutrition, suspecting that his food was poisoned.)

Gödel became deeply interested in relativity, and discovered (invented?) a possible universe that obeyed Einstein's equations and contained closed loops in time. Gödel's universe didn't much resemble our own – for one thing, it was not expanding. But this still raises the question: could there be a universe, resembling ours in that it contained galaxies and stars, where there were closed time-loops? If it took many billion years to go round the loop, then there would be no conflict with anything we experience, nor even (perhaps) with anything astronomers can observe. More recently, Richard Gott has proposed another kind of universe containing time-loops. Gott's universe is empty except for two infinite cosmic strings (see Chapter 11) in fast relative motion.

Our own universe would not allow time travel unless the requisite wormhole could be created; even then, we could never travel back to any era predating the wormhole's manufacture. Gödel's universe, and Gott's, would in contrast always have been so warped so as to allow time travel. Anyone traversing a time-loop must experience a history that closes up self-consistently. The philosophical problems thereby posed for the concept of free will would always have existed in Gödel and Gott's universes.

What attitude should we take to such scenarios? Perhaps their paradoxical character is pointing us towards some new and more restrictive law which rules out any universes with this unpalatable feature. (This recalls the attitude of those who want to throw out the solutions of Einstein's equations that aren't compatible with Mach's principle. See Chapter 12.) On the other hand, one could merely note that these closed loops lead to no obvious absurdity in a physical world where 'time police' ensure that everything closes up self-consistently each time one goes round the loop. Nevertheless, most physicists would probably share the view, maybe just a conservative prejudice, that some deep physical principle that we don't yet understand safeguards nature against anything that violates the normal order of cause and effect.

FLUCTUATIONS, ETERNAL RETURN, AND DUPLICATION

We have already noted that cosmic expansion defines an arrow of time. In contrast, a closed, finite, permanently isolated system in which gravity did not act would settle down into an equilibrium: there would be no overall change or trend to single out a particular arrow for time. The great 19th-century physicist Ludwig Boltzmann considered how the cosmic scene we actually see could have emerged from such a universe. (This was before external galaxies were known to exist. Boltzmann's 'universe' was therefore an aggregate of stars no larger than our Milky Way.) He was driven to postulate that everything within range of our telescopes was an incredibly rare fluctuation in an eternal, infinite cosmos. Even in its own terms, this was an unsatisfactory theory. Our existence may require a fluctuation as large as the Solar System, but there would be no reason why the fluctuation should extend as far as astronomers can probe into space. Indeed Boltzmann should have concluded that his brain was receiving coordinated stimuli that gave the illusion of a coherent external world which didn't actually exist. This solipsistic perspective would be vastly less improbable than the emergence of the whole external world as a random fluctuation!

Other paradoxes that arise in 'closed systems' are readily resolved by the cosmic expansion. Early this century, Henri Poincaré pointed out that any 'closed system' returns (indeed, returns infinitely often) to its present state – this would happen to the entire observable universe if Boltzmann had been right. But the so-called 'Poincaré recurrence time' is *immensely* long, even compared with cosmological timescales: only microscopic systems would repeat themselves within 10 billion years.[5] But if our universe continued expanding for ever, might there not eventually be time for a complete recurrence? The answer would be 'yes' for a system of fixed size. In an expanding universe, however, the amount of material in causal connection with a given mass element increases indefinitely.

An infinite universe might contain 'duplicates' of ourselves, following exactly parallel evolution for 10 billion years. But they would lie far beyond our present observational horizon. Eventually, light from these 'duplicates' may reach us. But, even if their history had mimicked ours for 10 billion years, there would be no reason why it should continue to 'track' ours in the remote future. There would, by then, have been far more time for variety to develop. Systems whose entire history has paralleled our own may exist, but they would become ever more thinly spread – our nearest 'duplicate' lies ever further beyond our horizon.

NOTES

1 Some authors, for instance the versatile Russian physicist Vitaly Ginzburg, conjecture that some other barrier may intervene before we even get near the Planck time. There may be a fundamental lattice structure in space-time on a scale much larger than 10^{-43} seconds. All that we know from direct experiments is that this scale cannot exceed 10^{-26} seconds.

2 It is helpful to envisage a 'space-time diagram' where successive times are horizontal slices, later times being higher up, and the 'world lines' of objects in the universe point straight upwards. If we plot just two (rather than three) spatial dimensions, the big bang would be a smooth horizontal 'floor' at the bottom. But the 'crunch' would be a very jagged surface: some parts would reach downward: these correspond to parts of the universe that collapsed into black holes quite early on. The picture would look like a cave with a smooth floor (no stalagmites) but many stalactites drooping from the roof.

3 For clocks on or near the Earth, the situation is a bit more complicated by gravity, and by the Earth's rotation. That is why the effect is not the same if you fly from east to west.

4 An inflationary expansion, as discussed in Chapter 10, requires the content of the early universe to behave as though it had a tension.

5 For something the size of a human, the recurrence time is such a large number that, written down with 1000 zeros per page, it would require a stack of paperbacks with volume roughly that of the moon!

14

'COINCIDENCES' AND THE ECOLOGY OF UNIVERSES

> Universes . . . might have been botched and bungled throughout an eternity ere this system was struck out; much labour lost, many fruitless trials made, and a slow but continual improvement carried out during infinite ages in the art of world-making.
>
> David Hume
> (1779)

HOW CONSTANT ARE NATURE'S 'CONSTANTS'?

We should be astonished if atoms behaved differently elsewhere in the world, or from one year to the next. The presumption that physics is the same everywhere is so deeply rooted that we tend to forget that this universality is learned from experience. The basic laws seem the same not only everywhere on Earth, but also (so far as we can tell) in every part of our universe that we can observe. Distant galaxies contain oxygen, sodium and other atoms, emitting light of just the same colours as are revealed by laboratory spectra of those same atoms. (We must of course correct for the redshift, which changes all wavelengths in the same proportion.)

The entire physical world, not just atoms, but stars and people as well, is essentially determined by a few basic 'constants': the masses of some so-called elementary particles, and the strength of the forces – electric, nuclear and gravitational – that bind them together and govern their motions.

The numerical values of these quantities depend on what units they are measured in. But the statement that, for instance, one object is 10 times heavier than another is equally true whether we measure masses in grams or ounces. *Ratios*, whether between two masses or two prices, don't depend on our choice of measures or currencies, and therefore have more significance. One such ratio is that between the masses of a proton and an electron, two basic constituents of atoms: its value is 1836. The sizes of atoms are also important. These are determined by how closely the (negatively charged) electrons are held in orbit around the central nucleus (positively charged because of its constituent protons). This depends on how much small-scale 'fuzziness' is introduced by quantum effects. The basic number here, known as the fine 'structure constant', tells us how much bigger atoms have to be (because of the quantum uncertainty principle) than the electrons themselves.

The basic 'physical constants', like the atoms and electrons, are legacies from the early universe. They would have been 'imprinted' right back in the initial instants – maybe when our universe had been expanding for less than 10^{-36} seconds. Conditions were then so extreme that experiments offer no guidance. Any underlying principle that relates these apparently arbitrary numbers – if such a principle exists at all – will surely require new insights about how our universe began.

CHANGING FORCES?

Einstein believed that gravity was genuinely universal, and that its strength shouldn't change with time. But others have argued, contrariwise, that it would be rather surprising for physical laws to be *un*changing in a changing universe. Can we be sure whether the 'constants' have indeed *stayed* constant through the entire stretch of time – ten billion (10^{10}) years – over which our universe itself has changed?

The mathematical physicist Paul Dirac made his greatest scientific contributions, while still in his twenties, during the 'heroic age' of quantum theory between 1925 and 1930; but he later became interested in cosmology, and proposed in 1937 that gravity was weakening as the universe grew older. Any such change would be a tiny effect – no more than one part in 10^{10} per year – and neither common sense nor everyday experience rules this out. Dirac's idea could not be checked at the time, but exact measurements can now test it. If gravity were changing, the orbits of planets and space probes would be slightly affected. Planets would spiral outward as the Sun's gravitational grip weakened; the cumulative effect might be discerned even within a few years, if the orbits could be tracked accurately enough. Accurate monitoring of NASA's Viking probe to Mars showed that gravity cannot change by more than 3 parts in 10^{11} per year: this rules out Dirac's suggestion, though an even slower change (at, for instance, a tenth the rate he predicted) cannot be ruled out.

Had gravity been stronger in the past, the Sun's core would have been squeezed to higher pressure. It would have burned brighter (boiling the oceans of the young Earth) and consumed its nuclear fuel faster. The same would happen to all the stars in every galaxy. Distant galaxies, whose light set out towards us when the universe was younger, would therefore be brighter than we'd otherwise expect.[1] No such evidence has shown up – young galaxies indeed look different, but only to the extent that's expected because their stars are at an earlier evolutionary stage, and because the 'recycling' between gas and stars has progressed less far. The actual contraints on gravity implied by such observations are, however, less 'clean' than those inferred from orbits in the Solar System, which are more straightforward to calculate.

The most precise tests of all, however, come from pulsars – neutron stars whose spin-rate provides an accurate clock. Some pulsars are orbiting around a companion star. Because pulsars can be timed very precisely, these orbits can be measured as accurately as those of planets or space probes in our own Solar System. If gravity were getting weaker, a pulsar would tend to move outward in its orbit, and lag behind its predicted position. This technique now offers even stronger evidence on the constancy of gravity than we can get from orbits within our own Solar System. (High-precision measurements of pulsars in binary orbits have already been mentioned in Chapter 4.)

Quite apart from gravity, one might wonder about the other basic numbers that govern the microworld – for instance, the masses of protons and electrons, or the strengths of electrical and nuclear forces. Might these have been different in the past? Could they even differ slightly from place to place, so that the light from a distant galaxy billions of light-years away may be different from the light from a flame, or from the Sun?

When the light from galaxies and quasars is analysed, all spectral lines are redshifted by the same factor. Once the redshift is allowed for, these atoms and molecules, remote from us in space and time, seem to be identical with those in the lab. But if protons, now 1836 times heavier than electrons, had a different mass in the past, then the spectral lines of hydrogen would be slightly shifted *relative to* those from other atoms.[2]

Some of the strongest evidence on the values of these constants in the distant past, comes (surprisingly) not from astronomers observing distant objects, but from geologists inferring past conditions here on Earth. At the Oklo uranium mine in Gabon in West Africa, various elements and isotopes[3] are found in unusual proportions. Geologists infer that, for some reason, enriched uranium and sea water accumulated on this site to form a *natural nuclear reactor* which 'went critical' about 2 billion years ago. This reactor reveals the effects of radioactive decay in the remote past. The rare element sumarium is particularly interesting, because one of its isotopes has an anomalously high propensity to 'soak up' the neutrons that are produced by radioactivity; it would therefore not survive prolonged exposure in a reactor. The Oklo samples indeed show very little of this isotope. This means that the basic forces that hold nuclei together – the nuclear (or 'strong') force and the electromagnetic force – can hardly have changed at all over the last 2 billion years. Otherwise the sumarium isotope wouldn't, during most of its history, have had an especially large chance of being destroyed. This limit is remarkably stringent, because sumerium would lose its unusual properties even if its nucleus changed by only a few parts in a billion. There cannot have been even that much change over a timespan of 2 billion years. The electrical forces cannot have changed by more than a few parts in 10^{17} per year. Changes of subatomic forces are therefore even more constrained than those of gravity.

So none of the basic forces, nor the masses of subatomic particles, can have changed substantially, even in several billion years. Exceedingly small changes, of course, cannot be excluded. In 'superstring' theories (see Chapter 9), particle masses are related to structures in extra-spatial dimensions which are 'wrapped up'; these may change slightly in response to cosmic expansion. The key phenomena of superstring theories are on scales very much smaller than any known particle. The prospects of directly testing such theories are bleak, and so this indirect test provides an extra motivation for pursuing these delicate searches for tiny changes in basic physical quantities.

LARGE NUMBERS AND WEAK GRAVITY

Dirac had an interesting motive for conjecturing that gravity was getting weaker. He noted that gravitational and electric forces both obey the inverse square law. Therefore, the ratio of the strengths of the electrical and gravitational forces between (say) an electron and a proton is a fundamental number. This number is exceedingly large: about 10^{39}. Dirac was then surprised to find that the size of the observable universe (the Hubble radius) exceeds the size of a proton by a factor that was also around 10^{39}. He also estimated the number of atoms in the observable universe (this is an even rougher estimate) to be about 10^{78}, the square of the previous number. Being reluctant to treat the similarities as a coincidence, he conjectured that there must be some underlying linkage between these large numbers.[4]

But if the cosmic expansion started with a big bang, as Dirac believed, then the size of the universe, or Hubble radius, certainly increases as the universe gets older: it is essentially the speed of light multiplied by the time since the big bang. So one of Dirac's numbers, the ratio of the Hubble radius to the size of a proton, increases with time. If the large numbers were indeed interlinked, then the ratio of electrical and gravitational forces would also have to increase in step with the time since the big bang. Otherwise it would be a coincidence that two apparently unrelated large numbers were equal at the present time. He proposed, therefore, that the gravitational force weakened inversely with the age of our universe. (Dirac could equally well have argued, instead, that the electrical force was getting stronger, but he thought gravity was more likely to respond to the large-scale universe.)

As already mentioned, if gravity were indeed getting weaker, the planets (or any artificial orbiting objects) would gradually spiral outwards as the Sun's grasp weakened. Measurements of spacecraft trajectories are precise enough to exclude a change in gravity even a tenth as fast as Dirac conjectured. But, even before this evidence, there was good reason for querying the basis for his conjecture.

DICKE'S RESPONSE TO DIRAC

The 'coincidences' that impressed Dirac are actually illusory. They were put in a new perspective by Robert Dicke, the Princeton physicist already mentioned in Chapter 3, who predicted the cosmic background radiation – and who, with better luck, would have discovered it before Penzias and Wilson did. Dicke appreciated that there was one large number in physics – the number that arises when we compare gravity with the forces that govern the microworld of atoms and elementary particles. The ratio of the electrical force and the gravitational force between the two protons in a hydrogen molecule is about 10^{36} (this is 10^3 different from Dirac's number because it refers to the forces between two protons, rather than between a proton and an electron). So gravity is negligible, by an enormous margin, in single molecules. However, everything exerts a positive gravitational attraction on everything else: there is no cancellation of positive and negative charges as occurs for electrical forces. So gravity wins on sufficiently large scales; Dicke was the first person to point out clearly how the properties of stars depend on this large number.

Suppose you start with a single molecule, and then assemble progressively larger lumps containing 10, 100, 1000 atoms and so on. The 24th, containing 10^{24} atoms, would be the size of a sugar lump; the 40th would be the size of a mountain or small asteroid. The effect of gravity on each atom depends on the mass of the lump it belongs to, divided by the radius. It goes up in proportion to the total number of atoms, *N*, but down by their average distance from each other, which scales as the cube root of *N* if the density is constant. The net result is that the importance of gravity goes as $N/N^{1/3}$: in other words, as the two-thirds power of *N*. For each thousand-fold rise in *N*, gravity gains by a hundred. Despite its initial handicap, amounting to 36 powers of 10, gravitational force becomes dominant when more than about 10^{54} protons are packed together (36 being two-thirds of 54). This is about the mass of Jupiter. Anything larger becomes a star. It is because gravity is so weak that a typical star like the Sun contains as many as 10^{57} atoms. In any lesser aggregate, gravity could not squeeze the material to high enough central densities and pressures for nuclear fusion to occur.

In the same spirit, Dicke could estimate how long a star would live for. This is related to the time a photon takes to diffuse, or 'random-walk', out of the star, carrying away the star's internal heat. This time, proportional to the total number of steps in the random walk, goes as the *square* of the radius (for an object of given density), and therefore as the two-thirds power of the mass – in other words, it scales in just the same way as the gravitational binding energy. The time it takes a photon to escape from a star, a measure of the star's lifetime, is therefore 10^{36} times longer than light takes to cross a single atom.

For life like us to evolve, there must be time for early generations of stars to have evolved and died, to produce the chemical elements, and then time for the Sun to form and for evolution to take place on a planet around it. This takes several billion years. By thinking clearly about the microscopic processes going on in stars, Dicke realised that their lifetimes, like their masses, involved the ratio of electrical and gravitational forces. When our universe is as old as a star, its observed size, roughly measured by the distance light has been able to travel since it began, will be 10^{36} times larger than an atom. Since we are observing the universe not at a random epoch in its history, but *when its age is about that of a star* – when stars have had time to form and evolve, but haven't all yet died – we automatically find Dirac's 'coincidence' to be satisfied.

The size of our universe shouldn't surprise us: its extravagant scale is necessary to allow *enough time* for life to evolve on even one planet around one star in one galaxy.

This is an example of an 'anthropic' argument, which entails realising that the Copernican principle of cosmic modesty should not be taken too far. We are reluctant to assign ourselves a central position, but it may be equally unrealistic to deny that our situation in space and time is privileged in any sense. We are clearly not at a typical place in the universe: we are on a planet with special properties, orbiting around a stable star. Somewhat less trivially, we are observing the universe not at a random time, but at a time when the requirements for complex evolution can be met. What Dicke showed was that the similarity of the two large numbers, which Dirac regarded as so significant, was *automatically fulfilled* during the era in the universe when we could exist (and which would, indeed, be the most propitious era for any 'observer' that Dicke could readily envisage).

Dirac reacted to this argument, incidentally, by accepting that there would not be life before there were stars, but hoping that some form of life would continue after the stars had died – not an absurd aspiration (as noted in Chapter 12); but an ineffectual response to Dicke because we are manifestly a specific kind of life that *does* depend on the warmth of a star.

A BAYESIAN DEDUCTION

Dicke proposed his idea in 1961. He could, although he did not, have developed it into an argument favouring the big bang over the steady-state theory. (The latter was still taken seriously at that time.) There will always be a stage in the evolution of any big-bang universe when its age equals the age of a typical star. We shouldn't be surprised to find that we are observing such a universe at this particular stage – we are made from the debris of dead stars, but depend on the warmth from a star that is still burning. But in a steady-state universe the expansion timescale of the universe (the Hubble time) is *always the same*. The physics that determines the Hubble time, though unknown, would have had nothing to do with stars or their evolution. There is no a priori reason why, in a steady-state universe, the timescale for stellar evolution and the Hubble time should be anywhere near the same: the stellar lifetime could be much shorter, in which case nearly all the matter would be in dead stars or burned-out galaxies; or it could be much greater than the Hubble time, in which case only an exceptionally old galaxy would look like our own. So, the fact that the lifetime of solar-type stars is comparable to the Hubble time would be an unexpected and improbable-seeming coincidence in steady-state theory, whereas it is entirely natural in a big-bang model.

This line of reasoning has real evidential value whenever one is trying to compare the claims of two rival theories – in the above example, big bang (B) versus steady state (S). Your prior opinion would be reflected in the odds you would offer if you were betting on which theory would 'win'. Suppose some new evidence emerges that is entirely unsurprising if B is correct but would seem improbable or coincidental in S. Then, *whatever your previous odds were*, you would now shift them towards B. (And, even if S remained your favourite, you would not bet so strongly against B as before.) This method of scientific inference is known as 'Bayesian', after the Reverend Thomas Bayes, who developed it in the 18th century.

ARE PHYSICAL CONSTANTS 'FINE-TUNED' FOR LIFE?

Claims that our emergence required special conditions or 'fine tuning' have a long history. They trace back to the classic theological case for an intelligent (and even benign) creator. The most celebrated expositor of these 'arguments from design' was William Paley, whose book *Natural Theology*, published in 1802, introduced the famous analogy of the watch and the watchmaker. Paley's arguments, mainly from biology – his 'Evidences for the Existence and Attributes of the Deity collected from the Appearances of Nature' – retain little interest, even for theologians, in the post-Darwinian intellectual climate. The inanimate world, the cosmos, impressed Paley more by its vastness than by its 'design'. To quote him on this point, 'My opinion of astronomy has always been that it is not the best medium through which to prove the agency of an intelligent creator, but that this being proved, it shows beyond all other sciences the magnificence of his operations.'

In a more scientific spirit, Lawrence Henderson, a Harvard professor, wrote two important books in this vein early in this century: *The Fitness of the Environment*, published in 1913, and *The Order of Nature*, in 1917. He was impressed by the seemingly surprising properties of some ordinary inorganic substances, which allowed conditions especially propitious for life. For example, water is very unusual among liquids in that it *expands* when it freezes – ice floats, making it much harder for a pond to freeze solid. He noted similar 'anomalies' in other important molecules such as carbon dioxide. Henderson concluded that 'we are obliged to regard this collection of properties as in some intelligible sense a preparation for the process of planetary evolution. Therefore the properties of the elements must for the present be regarded as possessing a teleological character.'

Henderson's arguments pertain to basic physics and chemistry, and cannot be as readily discounted as those of Paley concerning the 'fitness' of animals and plants for their environment. Any complicated biological contrivance is the outcome of prolonged evolutionary selection, involving symbiosis with its surroundings; but the basic laws governing atoms and molecules are 'given', and nothing biological can react back on them to modify them.

Advances since Henderson's day have disclosed other apparent coincidences where a rather delicate balance seems to prevail. Most crucial of all, perhaps, is the balance in the nuclei of atoms between the two forces that control their constituent protons and neutrons: the electrical repulsion between protons, and the strong nuclear force between protons and neutrons. If nuclear forces were slightly *weaker*, no chemical elements other than hydrogen would be stable and there would be no nuclear energy to power stars. But, if the nuclear forces were slightly *stronger* than they actually are relative to electric forces, two protons could stick together so readily that ordinary hydrogen would not exist, and stars would evolve quite differently.

The fact that the electron weighs so little compared with the nuclei of atoms is also important. It is a prerequisite for molecules like DNA maintaining their precise and distinctive structures. Heisenberg's uncertainty principle implies an inevitable 'fuzziness' in the location of any particle; the uncertainty is less for heavier particles. In a molecule, the uncertainty in an atom's position is determined by the mass of its nucleus. The orbits of electrons around the nucleus are very much larger, since electrons are lighter. It is the electron mass that determines the overall size of atoms, and the spacing between the atoms in a molecule. Because protons are 1836 times heavier than electrons, atoms can be quite precisely located relative to their distances from their neighbours, so complex molecules can have well-defined shapes.

A neutron is heavier than a proton by 0.14 per cent – little more than one part in a thousand. But this difference, small though it is, is important because it exceeds the total mass of an electron. If electrons weren't so light, they would combine with protons to form neutrons, leaving no hydrogen. (It requires the extreme pressures inside a neutron star for this process to happen in our actual universe.)

A 'SPECIAL' UNIVERSE?

All the chemical elements – iron, carbon, oxygen and so on – of which we are made were synthesised from primordial hydrogen and helium inside stars that exploded before our Solar System formed. And the chains of nuclear reactions that transmuted the elements depended on further apparent accidents. The most impressive of these was Fred Hoyle's recognition, described in Chapter 1, that some fine-tuned features of carbon and oxygen, seeming accidents of nuclear physics, turn out crucial for the pervasiveness of carbon, and therefore for the course of cosmic evolution.[5]

Had Henderson been writing today, he would have scanned our environment on a grander galactic (or even cosmological) scale, and noted the cosmic as well as terrestrial 'accidents' on which our emergence seemed to hinge. Our cosmic environment had to offer the scale and stability to provide the backdrop to these events. The special properties of the carbon nucleus would surely have impressed him.

Henderson would also have noted other 'coincidences'. As well as those involving the nuclear and electric forces, important in allowing stable nuclei to exist, there are others that involve neutrinos and radioactivity. Neutrinos interact weakly and infrequently with other particles: most of those hitting the Earth go straight through.

Nuclear reactions involving the creation or annihilation of neutrinos are consequently slow and inefficient. Such reactions are, however, crucial in building up the chemical elements.

Neutrinos control helium production in the big bang. A helium nucleus is made from two protons and two neutrons; the synthesis occurs when the temperature has fallen to a billion degrees; the amount of helium depends on how many neutrons survive up to that time. Neutrinos tend to remove neutrons, converting them into protons. The more efficient this reaction is, the fewer neutrons survive, and the less helium emerges. It is something of a coincidence that the neutrino reactions lead to the proportion of helium emerging from the big bang being around 25 per cent, rather than either zero or 100 per cent.

Neutrinos are, moreover, important to element production for a second reason. When a massive star explodes as a supernova, its inner core collapses to colossal densities, and generates an intense burst of neutrinos; these diffuse outward, depositing their energy in the outer layers of the star, which then get blown off. If the coupling between neutrinos and ordinary atoms were much larger, the neutrinos would stay trapped in the core; on the other hand, if this coupling were even weaker than it actually is, neutrinos would all escape freely. In neither case could they be effective in triggering the supernova explosions that expel processed material back into interstellar space. The oxygen and other elements forged within early stars must have been explosively dispersed through interstellar space, enabling second-generation stars like our Sun to acquire the basic materials for planets and life.

So the formation of all the elements depends – for these two quite different reasons – on the neutrino interactions having a strength that is rather specifically pinned down, compared with the range (spanning many factors of 10) in which it could lie, a priori.

The larger-scale features of our universe – galaxies, clusters and superclusters – evolved from ripples or fluctuations in the early universe (see Chapter 7). Obviously galaxies couldn't form if the density of matter was too low – gravity could never overwhelm pressure in a universe that just contained radiation. But, even if there is enough matter, something else is needed as well: initial irregularities to 'seed' the condensation of galaxies. As the expansion proceeds, regions denser than average lagged behind, and eventually condensed out to form structures.

The 'height' of the ripples superimposed on the overall smoothness is described by the number *Q*, another important constant of nature. A much smoother universe, where *Q* was even smaller than 10^{-5}, would for ever remain dark and featureless: no galaxies, and no stars, could ever form. The fabric of a universe with *Q* much bigger than 10^{-5} would be highly irregular with structure stretching up to the scale of the horizon (rather than, as in ours, being restricted to about one per cent of that scale). As described in Chapter 7, the cosmic scene would be dominated by black holes rather than galaxies, and stars (even if they managed to form) would be buffeted too frequently to retain stable planetary systems.

It will be interesting to complement the computer models described in Chapter 7 with comparable simulations for universes with very different values of *Q*.

GRAVITY: THE WEAKER THE BETTER

In a universe where electric forces between nuclei were even a few per cent stronger compared with the strong nuclear forces, the periodic table would contain just one element, hydrogen, instead of 92. Is anything in our universe equally sensitive to the strength of gravity?

If gravity were just a few per cent different, nothing would be drastically changed. The gravitational force is, however, feebler than the forces governing the microworld by about 10^{36} – one followed by 36 zeros. In the perspective of such a huge number, we should contemplate a much bigger variation. Imagine, for example, a universe where this number had several fewer zeros. How drastically would it differ from our own universe?

A large, long-lived and stable universe depends quite essentially on the gravitational force being exceedingly weak. It is because gravity is so weak that a typical star like the Sun is so massive. In any lesser aggregate, gravity could not compete with pressure, nor squeeze the material to high enough central densities and pressures for nuclear fusion to occur.

Imagine, for instance, a universe where there were 10 fewer zeros in the number measuring the weakness of gravity – where it was 'only' 10^{26} rather than 10^{36} times weaker than the microphysical forces in a hydrogen atom – but that microphysics was unchanged. In this universe, atoms and molecules would behave just as in ours, but objects would not need to be so large before gravity became competitive with the other forces. The number of atoms needed to make a star (a gravitationally bound fusion reactor) depends on the three-halves power of this large number. In our imagined universe, therefore, stars would have 10^{-15} the sun's mass. Their lifetimes would be 10^{10} times shorter. Instead of living for 10 billion years, a typical star would live for about a year.

All structures in such a universe, including galaxies, would be scaled down. Instead of the stars being widely dispersed, they would be so densely packed that close encounters would be frequent. This would in itself preclude stable planetary systems, because the orbits would be disturbed by passing stars – something which (fortunately for our Earth) is unlikely to happen in our own Solar System.

We might define planets as objects that are too small to become stars, but yet large enough for their self-gravity to affect their shape (making them more or less round) and perhaps to retain an atmosphere. Planet masses would therefore be scaled down by 10^{15}, like those of stars. Irrespective of whether these planets could retain steady orbits, the strength of gravity would stunt the evolutionary potential on them.

On our Earth, the size of living creatures is limited by gravity. Galileo knew this. Imagine any animal at double its actual size. Its linear dimensions are scaled up by 2, its volume and weight goes up by 8; but the cross-sections of its legs go up only fourfold, and would be too weak to support it. It would need a redesign: larger creatures need thicker legs relative to their overall size; giants would need legs thicker than their bodies. Simple physics sets a crude limit to the size of animals on Earth; moreover, this maximum size scales with the strength of gravity. If gravity were 10^{10} stronger, the limiting mass of an animal (in the sense of the number of atoms it could contain) would be less by about 30 million – no animals could be much larger than insects, and even they would need thick legs to support them.

The (literally) crushing effect of strong gravity would restrict the scope of complex evolution on this hypothetical world. But more severe still is the limited time. Chemical and metabolic processes would not be speeded up. But the mini-Sun would burn faster, and would have exhausted its energy before even the first steps in organic evolution had got under way.[6]

The actual scaling laws are not straightforward; nevertheless the conditions for complex evolution would undoubtedly be less propitious if (leaving everything else unchanged) gravity were stronger. There would be fewer powers of 10 between astrophysical timescales and the basic microphysical timescales for physical or chemical reactions. Complex structures could not grow very large without themselves getting crushed by gravity. Although gravity always dominates on sufficiently large scales it is because it is actually so weak compared with other forces that very large and long-lived systems can exist. In any 'interesting' universe there must be at least one very big number.

HOW MANY DIMENSIONS?

Our imagination isn't overstretched by universes with short lives, or where the basic forces act with different strengths, or even with a different 'zoo' of fundamental particles. But it is harder to get our minds round universes whose space and time admit different numbers of dimensions. Our universe is best described as '3 + 1' dimensional; there are four dimensions, but one of them, time, is distinctively different from the other three. Time is special insofar as we seem to be dragged only one way in it ('forward'); in the other three, we can move in either direction (east or west, north or south, up or down). This form of space-time has special mathematical features. For instance, if an object is rotated in an arbitrary way, it takes three numbers – the same as the number of dimensions – to specify the rotation: two to fix the direction of the rotation axis, and one to specify the angle through which it rotates around that axis.

Gravity and electric forces obey an inverse square law *because* we live in a universe with three space dimensions. This is easiest to appreciate in terms of Faraday's concept of lines of force. A shell of radius r around a mass or charge has an area proportional to r^2; the force falls off as $1/r^2$ because at larger radii the lines of force are spread over a bigger area, and their effect is diluted. In four-dimensional space, the area of a 'sphere' would go as r^3, and the force would follow an inverse cube law.

Planets could not remain in orbit if gravity obeyed an inverse cube (or steeper) law: if a planet were slightly slowed down, it would then spiral into the Sun, rather than merely be perturbed into a slightly smaller orbit. It was the theologian Paley[7] who first noted the special stability of an inverse square law. This buttressed his arguments for divine providence; but he did not relate it to the number of dimensions of space.

Some new insight may eventually reveal that space-time with 3 + 1 dimensions, like ours, is the only possible one. But there currently seems nothing absurd about a universe with extra dimensions. According to superstring theories, the ultra-early universe had ten dimensions. The extra six would have rolled up and 'compactified', rather than expanding along with the others. Theorists can't yet tell us whether this compactification inevitably leads to our 3 + 1 dimensions. (Whether a universe could have more than one *time*-dimension is less straightforward. Certainly a language with more tenses would be needed to describe what happens in it!)

AN ENSEMBLE OF UNIVERSES?

The laws of microphysics were imprinted during an the ultra-early phase of cosmic expansion. So also were other key features of our universe, for instance the number of baryons that are eventually available to form stars and galaxies, the amount of dark matter whose gravity binds galaxies together, and the number *Q* which measures the deviations from complete smoothness.

The physical laws 'laid down' in the big bang seem to apply everywhere we can now observe. But though they are unchanging (or almost so), they seem rather specially adjusted. This could be a coincidence: I used to think so. But an enlarged cosmological perspective suggests an interpretation that seems compellingly convincing. There may be other universes – uncountably many of them – of which ours is just one. In the others, the laws and constants are different. But ours is not randomly selected. It belongs to the unusual subset that allow complexity and consciousness to develop. Once we accept this, the seemingly 'designed' or 'fine-tuned' features of our universe need occasion no surprise. The next chapter develops this line of thought.

NOTES

1 If gravity were slightly higher in the past, the inferred ages of stars would be lower. Such a hypothesis could ease the problem (Chapter 8) that stars seem as old as the universe. This idea has not been pursued. There is no other motivation for changes in *G*: indeed, the idea would force us to abandon the successes of Einstein's general relativity. Moreover, the law governing how *G* changes would need to be rather carefully contrived in order not to violate any other observations.

2 There would be interesting changes even within the hydrogen atom itself. The characteristic radiation that hydrogen emits in the radio band, at a wavelength of 21cm, depends in a different way on the electron mass, Planck's constant and so forth from the spectral features seen in the optical and ultraviolet. But radio and optical astronomers get the same answer when they determine a distant object's redshift.

3 An element is defined by its 'atomic number' in the periodic table: the number of protons in its nucleus (whose positive electric charges are cancelled by an equal number of electrons orbiting the nucleus). There may be variants of the same element with different numbers of neutrons in the nucleus. These variants are called 'isotopes'. Deuterium, for instance, with one proton plus one neutron, is a heavy isotope of hydrogen (just one proton). For a description of the Oklo reactor, see the article by M. Maurette in *Annual Reviews of Nuclear Science*, vol. 26, p319 (1976).

4 Dirac made his suggestion at a time when his Cambridge contemporary Eddington, already celebrated for his classic and durable work on relativity and stellar structure, had moved on to develop an elaborate numerological 'fundamental theory', according to which our universe was closed and finite, and there were, as he famously asserted, '15,747,724, 136, 275, 002, 577, 605, 653, 961, 181, 555, 468, 044, 717, 914, 527, 116, 709, 366,231,425,076,185,631,031,296 protons in the universe and the same number of electrons'. (This number is actually $2^{256} \times 136$.) No living scientist believes this, and hardly any have thought it worth making the effort to fathom Eddington's reasoning.

5 It was especially impressive that Hoyle *predicted* the key property of the carbon nucleus, with one per cent precision, before it was experimentally measured. But the apparent 'fine tuning' would be equally noteworthy; even if the carbon nucleus had already been probed before anyone started to explore the particular reactions involved in synthesising the chemical elements. A prediction is to the credit of the person who predicts it; but the import of a scientific 'coincidence' is independent of the order in which the pieces fell into place.

6 The details sketched here are simplified: other features, like the surface temperatures of stars, scale in a more complicated way. In the small-scale

high-gravity universe, the prospects for complex evolution would be better around a star that was the analogue not of our Sun (which would live for only one year) but of a star only just above the threshold for hydrogen burning. Such stars in our universe may last for 10^{13} years and radiate in the infrared; in the hypothetical 'speeded-up' universe they would last for 1000 years, and have rather hotter surfaces.

7 A sophisticated interest in scientific matters was by no means rare among clergy of Paley's time. Mathematics was routinely studied by academically ambitious students at Cambridge. The examination was highly competitive. Students who achieved what would now be called first-class honours in mathematics were termed 'Wranglers'. The student who came top of the list was called Senior Wrangler. Paley was Senior Wrangler in 1764; he would have been highly proficient in calculating orbits, and other consequences of Newton's laws – more so, indeed, than present-day students whose syllabus ranges more widely over modern topics. Paley also studied Greek and Latin, as well as theology. Country vicarages were a comfortable refuge for academics who at that time couldn't remain at Oxford or Cambridge Colleges if they got married. Thomas Bayes, the pioneer statistician, was a clergyman; and John Michell, whose 18th-century speculations about black holes were mentioned in Chapter 5, was vicar of the parish of Thornhill, in Yorkshire.

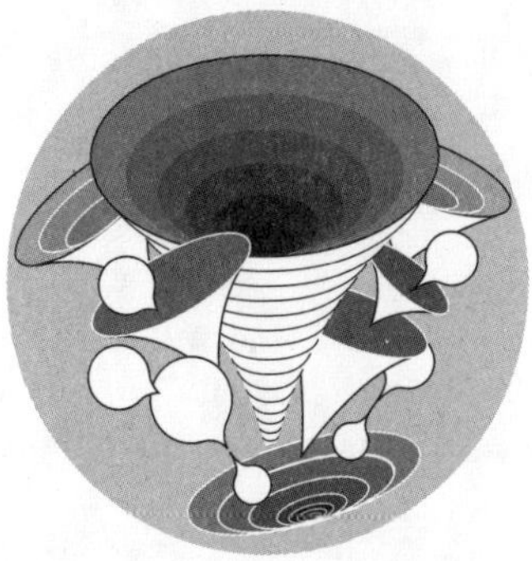

15

ANTHROPIC REASONING – PRINCIPLED AND UNPRINCIPLED

The universe knew we were coming.

Freeman Dyson

OUR CARBON-BASED BIOSPHERE HAS slowly evolved on a planet orbiting a stable star. Given that fact, some features of our universe, some constraints on the physical laws that govern it, follow quite straightforwardly. Various striking coincidences highlighted throughout this book acquire a different perspective once we recognise that we couldn't have evolved if they hadn't been fulfilled. But they still, I believe, have a deep significance: they offer clues to what the cosmos is like on even vaster scales than we can yet (or perhaps can ever) probe.

Our universe has some features – being long-lived, stable and far from thermal equilibrium, for instance – that are prerequisites for our existence; moreover, our emergence, as we have seen, depended crucially on apparent 'fine tuning' of the basic physical constants: the strengths of the fundamental forces, the masses of elementary particles and so forth.

There are various ways one can react. The most robustly dismissive attitude is that the physical constants must have *some* values, so we have no reason to be surprised at any particular value rather than another. Sciama gives a *reductio ad absurdum* of this viewpoint. Suppose you walk into a room and find a million cards laid out on an enormous table. Suppose you find that these cards are numbered progressively 1, 2, 3 and so on up to 1 000 000. Would you think they had been laid out at random, on the grounds that any particular ordering would have the same probability as any other? Obviously you would not.

This judgement doesn't depend on our knowledge that people may have reasons for dealing cards in a special order: there is an objective mathematical basis for saying this particular ordering is special. It is definable by a very short set of instructions: 'start with one, and then add one each time'. In contrast, a long and complicated program would be needed to tell a computer how to print out most sequences – indeed a purely random sequence can't be encapsulated by any program shorter than itself (that is, essentially, what 'random' means). Obviously the order implied by the cosmic coincidences doesn't stare us quite so plainly in the face as the ordering of cards. But this example counters the dismissive view that, whatever pattern we find, we should just accept it as chance rather than seeking an explanation.

A more reasonable reaction to the coincidences is to invoke a kind of 'selection effect'. Fishermen aren't surprised (to use an old metaphor of Eddington's) to catch no fish smaller than the holes in their nets. Nor are optical astronomers surprised that the objects they detect are very hot (since otherwise they wouldn't be shining). Likewise it may seem irrational to be surprised that our universe has any particular property if we wouldn't exist otherwise.

But even that doesn't seem quite enough. To say that we wouldn't be here if things were otherwise need not quench our curiosity and surprise that our universe is as it is. The Canadian philosopher John Leslie has given a nice analogy. Suppose you are facing execution by a fifty-man firing squad. The bullets are fired, and you find that all have missed their target. Had they not done so, you would not survive to ponder the matter. But, realising you were alive, you would legitimately be perplexed and wonder why.

It seems noteworthy, at the very least, that the physical laws governing our universe have allowed so much interesting complexity to emerge in it, especially as we can so readily imagine stillborn universes where nothing could evolve. If a 'cosmic being' turned knobs to vary the constants of physics and constructed a whole ensemble of universes, then clearly only one would be like our own, and we wouldn't feel 'at home' in most of them – that much is obvious. However, what is less trivial, and may be deeply significant, is that only a narrow range of hypothetical universes would allow *any* complexity to emerge.

This line of thought has a long lineage: its basis is often called the 'anthropic principle'. This name is an unfortunate one. 'Principles' in cosmology have often connoted assumptions unsupported by evidence, but without which the subject can make no progress.[1] The use of such phrases has hampered cosmologists in getting their subject accepted as an empirical one. I prefer the less pretentious phrase 'anthropic reasoning'.

The modern interest in anthropic reasoning was triggered by Brandon Carter, another student of Sciama, and one of my exact contemporaries at Cambridge. He was the first person to understand fully the internal structure and significance of the rotating black holes described by Roy Kerr's famous solution of Einstein's equations (see Chapter 5); he went on to contribute further ideas on black holes and (latterly) on cosmic strings. In 1970, Carter wrote a long manuscript discussing the 'coincidences' in the values of the basic physical constants, and pointing out some new ones – for instance, the formation of planetary systems around stars may require a particular relationship between the strength of gravity and various numbers derived from atomic physics. This was never published, though a shorter version appeared in print several years later. Carter's manuscript was, however, widely discussed, and he has been an influential participant in the ensuing debates.

Carter distinguished between 'weak' and 'strong' anthropic reasoning. What he called the 'weak anthropic principle' was essentially an allowance for observational selection: we shouldn't take Copernican modesty too far, but had to accept that creatures like us couldn't view the universe at all points in space and time, so that our perspective was bound to be in some sense special. Even this 'weak principle' allows some inferences which are interesting and by no means obvious. A good example was Dicke's demonstration (described in the previous chapter) that Dirac was wrong to be puzzled about the coincidences between two large numbers – one measuring the size of our observable universe, the other the strength of gravity. Dicke pointed out that we were not observing the universe at a random time in its history: we were living in an era when some but not all stars had died and straightforward stellar physics demands that Dirac's 'coincidence' automatically holds during that era.

Carter's '*strong* anthropic principle' was more controversial and speculative: this is the idea that the fundamental laws in any universe must actually be such as to permit observers to exist. Such a claim has teleological overtones; few have ever taken it seriously. It does, however, find an echo in some interpretations of quantum theory.

THE 'PARTICIPATORY' VIEW

Niels Bohr averred that anyone who was *not* astonished by quantum theory hadn't fully taken it in. Philosophical debates about this theory remain as lively as ever. The most common viewpoint, the so-called 'Copenhagen interpretation' espoused by Bohr and his followers, envisages that every system is governed by a 'wave function', whose behaviour is exactly determined by Schrödinger's famous equation until an observer chooses to make a particular measurement. This leads to what is called 'collapse of the wave function'. Information *complementary* to what is actually measured (for instance, the exact speed of a particle if we measure its position, or vice versa) is then irretrievably lost.

Quantum theory undeniably works; most physicists apply it confidently but unreflectively. As John Polkinghorne has put it, 'the average quantum mechanic is no more philosophical than the average motor mechanic'. But the 'Copenhagen interpretation' makes many of us uneasy. It implies a sharp and artificial-seeming disjunction between the object being measured and the observer (or experimenter); on the other hand everything – observer and experiment alike – should equally be subject to quantum laws.

Some have exalted the observer's role even further, claiming that an observation is needed to 'bring the world into being'. Universes would then be 'real' only if they contained observers: the status of a universe would depend fundamentally on whether it could harbour a conscious observer of any kind (whether or not it resembles carbon-based life). This 'participatory' view has been espoused by John Wheeler, who wrote,

> The system of shared experience which we call the world is viewed as building itself out of elementary quantum phenomena, elementary acts of observer-participancy. In other words, the questions that the participants put – and the answers they get – by their observing devices, plus their communication of their findings, take part in creating the impressions which we call the system: that whole great system which to a superficial look is time and space, particles and fields.

The 'participatory universe' seems hard to accept – hard, even, to take seriously. What sort of observer must be invoked to 'bring the universe into being': a mouse? a human? or a PhD physicist? I'm reluctant, however, to be too dismissive, because Wheeler has earned the right to be heeded. Early in his career, back in the 1930s, he collaborated with Niels Bohr on the theory of nuclear fission. Richard Feynman was one of his first PhD students (graduating in 1942) and Wheeler has ever since been an inspiration to successive generations of physicists. In the 1950s, he explored what quantum uncertainty might imply for the nature of space and time. He invented the concept of 'space-time foam' – the idea that on the Planck scale space and time themselves, and even the dimensionality of the world, fluctuate randomly. He introduced 'geons' – hypothetical entities that behave like particles, but are composed of nothing but curved empty space. Wheeler's bold conjectures prefigured the current ideas of chaotic cosmology and wormholes at the Planck time. He coined the term 'black hole' and was the primary inspiration behind the resurgence of American research in relativity in the 1960s. His gaze is still focused on the conceptual frontiers of cosmology – the 'flaming ramparts of the world', as he calls them.

The paradoxes of quantum mechanics, and the nature of consciousness, are manifestly two of the deepest mysteries of all. It is striking that John Wheeler and Roger Penrose, the most original and influential living theorists about space and time, have both, in their later years, advocated the dissident view that these mysteries are linked.

Whatever one thinks about Wheeler's concept of quantum mechanics, it shifts the perspective of anthropic arguments by making them less anthropocentric. Rather than asking what conditions were necessary for *our* evolution, we can ask what distinguishes a 'stillborn' universe from one that would be 'cognisable', in the sense that it permitted some kind of conscious entity or 'observer' to evolve within it. One must not be too anthropomorphic nor too restrictive in envisaging the requirements for the emergence of a conscious observer. Maybe neither stars, nor even atoms, are absolutely necessary. Science-fiction writers have familiarised us with many bizarre alternatives. However, some departure from thermal equilibrium would seem essential. A universe that recollapsed too promptly to a big crunch would, for instance, never become 'cognisable' during its brief existence. A force like gravity may also be essential, in order that structures should emerge (as described in Chapter 7); but it must be weak and slow-acting, as in our own universe.

Another approach to quantum mechanics is the 'many-worlds' theory, proposed by Hugh Everitt in the 1950s. The basic idea here is presaged in Olaf Stapledon's classic work of science fiction *Star Maker*:

> Whenever a creature was faced with several possible courses of action, it took them all, thereby creating many . . . distinct histories of the cosmos. Since in every evolutionary sequence of the cosmos there were many creatures and each was constantly faced with many possible courses, and the combinations of all their courses were innumerable, an infinity of distinct universes exfoliated from every moment of every temporal sequence.

The 'many-worlds' approach envisages our entire universe as a single quantum system. It appeals especially to cosmologists, for whom the Copenhagen picture is plainly unsatisfactory, since there cannot then be an observer external to the system when that 'system' is the entire universe. A newer variant due to David Deutsch replaces the idea of branching universes by an infinite ensemble of universes, evolving in parallel and displaying greater variety as time goes on.

ANTHROPIC SELECTION OF UNIVERSES

The physical constants seem to have fixed values throughout our universe. Maybe they could be different in others. Brandon Carter and I first heard such speculations floated in some public lectures in Cambridge in the early 1960s, given by Charles Pantin, a biology professor. He said that:

> the properties of the material universe are uniquely suitable for the evolution of living creatures. If we could know that our own universe was only one of an indefinite number with varying properties we could perhaps invoke a solution analogous to the principle of natural selection, that only in certain universes, which happen to include ours, are the conditions suitable for the existence of life, and unless that condition is fulfilled there will be no observers to note the fact.

If there were nothing beyond our universe, its properties indeed seem fine-tuned, or even providential. But suppose Pantin's conjecture could be fleshed out – suppose there really were other universes. If the 'constants' took different values in each, there would then be no need for surprise that some universes allowed creatures like us to exist. And we obviously find ourselves in one of that particular subset. If you go to a clothes shop with a large stock, it isn't surprising to find a suit that fits you.

Scientists commonly pay obeisance to 'Ockham's razor' – the celebrated injunction by William of Ockham, in the early 14th century, which (translated from his Latin) means 'don't multiply entities more than is absolutely necessary'. Nothing, perhaps, could seem to violate this more drastically than postulating an infinite array of universes! Nor does it, at first sight, seem properly 'scientific' to invoke regions that are unobservable, and perhaps always will be. The concept of an ensemble of universes of which ours is just one member (and not necessarily a typical one) is, needless to say, not yet in sharp theoretical focus. But it helps to explain basic (and previously mysterious) features of our universe, such as why it is so big, and why it is expanding. In the broader perspective of a 'multiverse', anthropic reasoning acquires genuine explanatory force.

A MULTIVERSE?

The 'many-worlds' version of quantum mechanics offers one approach to the 'multiverse' concept. The idea of 'eternal inflation' described in Chapter 10, though still very speculative, suggests another context in which other universes could exist.

The fundamental forces – gravity, nuclear, and electromagnetic – all 'froze out' with their present values and distinctive properties as our universe cooled down. Likewise the masses of the elementary particles. When the inflationary era ended, space itself (the 'vacuum') underwent a drastic change. As Chapter 10 describes, inflation may lead to separate universes – separate domains within a multiverse – which cooled down differently, ending up governed by different laws.

Complex evolution would occur only in 'oases' where the constants had propitious values. Our oasis must then be at least 10 billion light-years across because the physical laws seem the same everywhere we can observe. But the 'desert' beyond it would come into view in the remote future, when, maybe 10^{12} years or more from now, light from the edges of our domain has had time to reach us. This time delay is, to be sure, an impediment to empirical tests but such statements, though plainly not yet testable, are certainly not vacuous: they are akin to the conjectures of early explorers about what lurked beyond the frontiers of the then-known world, or Barrow and Tipler's speculation that 'the universe will recollapse 10^{15} years from now'.

The other universes may even be completely disjoint from ours, so that they will never come within the horizon of our remotest descendants. We may be part of an infinite and eternal multiverse within which new domains 'sprout' into universes whose horizons never overlap – ironically, the steady-state concept can then be revived, but applied to the multiverse rather than its constituent universes. New universes could be triggered by the big crunch (if our own universe eventually recollapses), or even within black holes. Alex Vilenkin prefers to envisage the universes as 'nucleating' separately and spontaneously, each thereafter inflating into its own separate space.

What these speculative viewpoints have in common is that all envisage our big bang as one event in a grander structure; the entire history of our universe is just an episode in the infinite multiverse. Earlier chapters have given a flavour of what other universes might be like.

The multiverse could encompass all possible values of fundamental constants, as well as universes that follow life cycles of very different durations: some, like ours, may expand for much more than 10 billion years; others may be 'stillborn' because they recollapse after a brief existence, or because the physical laws governing them aren't rich enough to permit complex consequences. In some there could be no gravity: or gravity could be overwhelmed by the repulsive effect of a cosmological constant (lambda), as it would have been during the early inflation phase of our own universe. In others, gravity could be so strong that it crushes anything large enough to evolve into a complex organism. Some could always be so dense that everything stayed close to equilibrium, with the same temperature everywere. Some could even have different numbers of dimensions from our own.

Even a universe that was, like ours, long-lived and stable, could contain just inert particles of dark matter, either because the physics precludes ordinary atoms from ever existing, or because atoms all annihilate with exactly equal numbers of antiatoms. Even if protons and hydrogen atoms exist, the nuclear forces may not be strong enough to hold the nuclei of heavy atoms together: there would then be no periodic table, and no chemistry.

SMOLIN'S SPECULATION

Natural selection of 'favoured' universes seems the stuff of science fiction. However the American cosmologist Lee Smolin conjectures that the multiverse could display the effects of heredity and selection. When a black hole collapses, he speculates that another universe sprouts from its interior, creating a new expanse of space and time disjoint from our own. Small universes, in which there was too little space or time to form many black holes, would not leave many progeny. Nor, he argues, would even a large universe if its physics prohibited stars from ever terminating as black holes.

Smolin then adds a new twist – the physical laws governing the daughter universe may differ from those in its parent, *but only slightly*. Since the number of progeny a universe has depends on the laws prevailing within it, there is a selection pressure. Many generations, or many iterations, would lead to a 'takeover' by the universes that generate the most numerous progeny. These would be the ones governed by laws that allowed the largest number of black holes to form.

The mechanisms that might 'imprint' the basic laws and constants in a new universe are obviously far beyond anything that we understand. But Smolin's idea, though far out on the speculative fringe, has the virtue of making a testable prediction: if our own universe is an outcome of such 'selection', it should be maximally efficient at spawning new universes – in other words, black holes must form more readily in it than they would in a different universe. This prediction can be addressed and checked by astronomers, using observations and physical ideas that are relatively conventional.

If we imagine tinkering with our universe by changing the electron mass, the strength of gravity and other variables, then any change, any 'turning of the knobs' from their actual setting, should, if Smolin is right, either make the universe smaller or reduce the propensity of stars to terminate as black holes. So to test his conjecture we need to know, for instance, how an adjustment of the constants would affect the chance that a heavy star ends up as a black hole, rather than throwing off so much mass during its lifetime that it can end up as a white dwarf (below Chandrasekhar's limiting mass, described in Chapter 4) or a neutron star. We also need to know whether the same adjustment would alter the processes of star formation, perhaps tilting the balance between high-mass stars (which might end as black holes) and low-mass stars (which will end as white dwarfs).

Star formation is something that astrophysicists will have to understand better before they can model the 'ecology' and recycling processes within even our own Milky Way, and before they can interpret the pictures of remote newly forming galaxies that the Space Telescope is giving us. Protostars are now condensing within dense clouds of gas and dust in, for example, the Orion Nebula, only 1500 light-years away. The detailed flow patterns that lead to protostars are hard to compute, but plainly depend on complicated interactions between gravity and pressure (which depends on the gas temperature); magnetic fields are also important; so is dust mixed with the gas, which emits infrared radiation and helps the clouds to cool and reduce their pressure. But how would stars have formed in a young galaxy at an earlier cosmic epoch? There would then have been no chemical elements other than hydrogen and helium (and therefore nothing that could make dust); probably no magnetic fields; and the primordial radiation, now at 2.7 degrees, would have been warmer, and might stop the clouds getting as cold as they are today.

Astrophysicists are still flummoxed about these questions, and this lack of understanding impedes all attempts to understand how galaxies evolve. It also impedes any test of Smolin's speculations. We can be confident that many black holes are indeed forming via stellar deaths or via the accumulation of huge masses in galactic centres. Our universe is therefore certainly not 'sterile' by Smolin's criterion. But, as to whether it is anywhere near 'optimal' for black-hole formation, we must suspend judgement.

Suppose, for instance, that a hypothetical universe was like ours except that the electric repulsive force within nuclei was slightly stronger, so they didn't fuse together to release energy and build up the elements of the periodic table. Stars would then evolve much more quickly to their final state. Low-mass stars would still become white dwarfs. But heavy stars would be more likely than in our actual universe to become black holes, because there would be no chance that nuclear energy could blow off enough material to bring them below the Chandrasekhar limit before they died. At first sight, this seems a straight disproof of Smolin. However, his conjecture could be salvaged if stars formed so differently in a medium of pure hydrogen that none were massive enough to become black holes and create progeny.

Quite apart from its relevance to other universes, astrophysicists would like to settle this issue because the first stars that formed in our expanding universe would have been made of hydrogen and helium; if these were indeed mainly of low mass, they would be candidates for baryonic dark matter (see Chapter 6). To elucidate it requires a better understanding of 'conventional' astronomical processes – an understanding that I hope we'll have within a few years – rather than exotic new concepts.[2]

Wheeler and Guth have speculated about artificial creation of small black holes by implosion mechanisms, which could be the seeds of new universes. Of course, anthropic arguments are superfluous if we live in a purposely created 'daughter universe'; Paley's old 'arguments from design' might then make a comeback.

ASSESSING ANTHROPIC ARGUMENTS

Where do anthropic arguments stand today? When they were first widely discussed, in the 1970s, many people were dismissive. Such arguments plainly didn't offer scientific explanations in the proper sense: at best they seemed a stop-gap, allaying our curiosity about phenomena for which genuine physical explanations were being sought, but had yet to be found. The world would, for instance, be very different if the nuclear or electric forces were somewhat altered, but one still hopes eventually to relate these forces via some mathematical formula just as, more than a century ago, James Clerk Maxwell related electrical and magnetic forces and the speed of light. Likewise, Salam and Weinberg took a further step in incorporating weak neutrino forces into an integrated scheme. By extension, a more comprehensive theory may eventually unify all the fundamental forces.

Rather than having to be measured experimentally, the physical forces and constants may one day be mathematically calculable – just as a circle's circumference can be calculated from its diameter, though not quite so easily. Weinberg says, 'I certainly wouldn't give up attempts to make the anthropic principle unnecessary by finding a theoretical basis for the value of all the constants. It's worth trying, and we have to assume that we shall succeed, otherwise we surely shall fail.' So it would be a pity if theoretical physicists took anthropic ideas too seriously, as it might diminish their motivation for seeking unified theories!

The physicist Heinz Pagels was especially dismissive:[3]

> Physicists and cosmologists who appeal to anthropic reasoning seem to me to be gratuitously abandoning the successful programme of conventional physical science of understanding the quantitative properties of our universe on the basis of universal physical laws. Perhaps their exasperation and frustration . . . has gotten the better of them. The influence of the anthropic principle on the development of contemporary cosmological models has been sterile. It has explained nothing, and it has even had a

> negative influence, as evidenced by the fact that the values of certain constants ... for which anthropic reasoning was once invoked as an explanation can now be explained by new physical laws . . . I would opt for rejecting the anthropic principle as needless clutter in the conceptual repertoire of science.

This goes too far in disparaging all anthropic reasoning. After all, in its weak form – the realisation that our existence, as observers, in itself has implications for what our cosmic environment must be like – little more is involved than the routine attitude of an observer who allows for the limitations of his search techniques and equipment. Saying that our cosmic environment is 'well tuned' for our sort of life is not just a tautology. It yields genuine understanding: Dicke's explanation of Dirac's alleged 'coincidences' (see Chapter 14) is an example. There is nothing very controversial about the claim that another universe could be 'malajusted' for life – for instance, its physical constants might preclude any chemical elements other than hydrogen. Universes that stay close to thermal equilibrium, which exist for too short a time, or (more radically) have only two spatial dimensions, would surely be less propitious than ours for complex evolution of any kind at all.

SCOPE AND LIMITS OF A 'FINAL THEORY'

The status and scope of anthropic arguments, in the long run, will depend on the character of the (still quite unknown) physical laws at the very deepest level. Weinberg hopes that a 'final theory' exists, and that we may some day discover it. Such a theory may uniquely fix the basic laws governing our universe: they could all be computed from some fundamental equations.

If the physical constants were indeed uniquely fixed by a 'final theory', it would then be a brute fact that these universal numbers happened to lie in the narrowly restricted range that permitted complexity and consciousness to emerge. The potentialities implicit in the fundamental equations – all the intricate structures in our universe – may astonish us, but this reaction would be akin to the surprise mathematicians must sometimes feel when vastly elaborate deductions follow from innocuous-looking axioms or postulates.

Consider, for instance, the Mandelbrot set. The instructions for drawing this astonishing pattern can be written in just a few lines, but it discloses layer upon layer of varied structure however much we magnify it. Anyone who has learned coordinate geometry (where the coordinates x and y denote distances along the two axes) can visualise that $x^2 + y^2 < 1$ is the inside of a circle; a real superintelligence would be needed immediately to visualise the shape of the Mandelbrot set just from reading its 'formula'! Similarly, latent in the succinct equations of a 'final theory' could be everything that has emerged in our universe, as it cooled from the initial big bang to the diffuse low-energy world we inhabit.

But the quest for a unique theory may be doomed to fail: the particles and forces in our universe could be inherently arbitrary. The way a metal bar responds to pressure or bending is important to engineers, and there are detailed tables giving the 'elastic constants' for different materials. But, to the physicist, the bulk properties of solids are *secondary* quantities determined by the underlying atomic structure. Likewise, what we call the fundamental constants – the numbers that matter to physicists – may be *secondary consequences* of the 'final theory', rather than direct manifestations of its deepest and most fundamental level. The multiverse may be governed by some unified theory, but each universe may cool down in a fashion that has 'accidental' features, ending up governed by different laws (and with different physical constants) from other members of the ensemble.

It could turn out that some constants are genuinely universal, even though others are not. Maybe those governing the microworld are unique throughout the multiverse (or at least in all universes with three spatial dimensions), and are therefore 'brute facts', but the constants important for cosmology differ from one universe to the next. Universes could then have distinctive values of omega (which fixes the density and how long their 'cycle' lasts if they recollapse), lambda (which measures the energy latent in empty space) and Q (which measures how smooth a universe is, and so determines what structures emerge in it).

If some physical constants are arbitrary, then anthropic arguments can properly be deployed to account for the values they actually take in our universe – indeed this would be *the only way* to understand why they didn't have values that were very different.

Any final theory is still such a distant goal that we cannot guess which features (if any) are unique throughout the multiverse, and which are 'accidents' of the way our universe cooled down. We therefore cannot yet assess how far our universe can be explained 'anthropically'. However, just as Smolin's speculative conjectures about natural selection of universes make 'predictions' that can be verified or refuted by quite conventional astrophysicists, so we may be able to fathom the nature of the final theory even before we know its specific details.

Suppose that the important physical constants take different values in other universes. Their measured values may not be typical of the entire multiverse: our universe must, as we have seen, have been special and atypical to permit our existence. But it ought not to be *unduly* special: this realisation allows us to guess which constants are 'accidents', and which are determined by the laws of a 'final theory' that prevail throughout the multiverse.[4]

Consider, for instance, the cosmological constant lambda, a repulsive force that accelerates the cosmic expansion. Will the 'final theory' offer some deep reason why this has to be exactly zero? Or is its value an 'accident'? We just don't know. Anthropic constraints rule out very large values of lambda because an unduly fierce cosmic repulsion would prevent galaxies from forming. Very large *negative* values would also be excluded, because the universe would then recollapse after a cycle too short to have allowed stars to evolve.

Lambda in our own universe may not be zero: as described in Chapter 8, the age estimates of stars could be more readily reconciled with a short Hubble time if the cosmic expansion were accelerating rather than slowing down. The lambda that is favoured (tentatively) is about ten times lower than the anthropically allowed maximum – no galaxies could form in a universe where lambda was more than ten times larger than the value some observers actually prefer. Thus lambda may not be enormously less than its allowed maximum – our universe would not, in respect of lambda, have been 'tuned' much more precisely than our presence requires.

On the other hand, if improved observations were to reveal that lambda had *no* discernible effect, implying that it was perhaps 1000 times smaller than anthropic constraints required, then our universe would seem even more special, in respect of lambda, than it need be. We might then suspect that the 'final theory' mandated an exactly zero lambda throughout the multiverse.

Even those theorists who hope that most constants will be uniquely pinned down are less optimistic about explaining lambda. Steven Weinberg, for instance, suspects that lambda may be the only number that is anthropically selected rather than being directly fixed by the final theory.

By applying similar arguments to other constants, we can infer what features are unique and uniform throughout the multiverse, and what variety the separate universes can display. Can *all* the microphysical constants and particle masses be different in other universes? And what about gravity? It could even – as Smolin conjectures – be natural selection, not a mere accident, that our universe (that is, the part of space-time we can observe) has the particular values of the physical constants that we measure.

OUR UNIVERSE AND OTHERS

Most universes would be less propitious for complex evolution than ours, but not necessarily all. We cannot conceive what structures might emerge in the distant future of our universe. Still less, therefore, can we envisage what might happen in a universe where there were more forces than the four we are familiar with, or where the number of dimensions was larger. Our universe could be 'impoverished' compared with some others, which could harbour vastly richer structures, and potentialities beyond our imaginings.

By mapping and exploring our universe, using all the techniques of astronomy, we are coming to understand – to a degree that even a decade ago would have seemed astonishing – our cosmic habitat, the laws that govern it, and how it evolved from its formative initial instants. But, even more remarkably, we have intimations of other universes, and can perhaps deduce something about them. We can infer the scope and limits of a final theory even if we are still far from reaching it – even if, indeed, it eludes our intellectual grasp for ever.

Chandrasekhar thought as long and intensely about our universe as anyone since Einstein. In one of his last lectures, he concluded:

> The pursuit of science has often been compared to the scaling of mountains, high and not so high. But who amongst us can hope, even in imagination, to scale the Everest and reach its summit

when the sky is blue and the air is still, and in the stillness of the air survey the entire Himalayan range in the dazzling white of the snow stretching to infinity? None of us can hope for a comparable vision of nature and of the universe around us. But there is nothing mean or lowly in standing in the valley below and awaiting the Sun to rise over Kinchinjunga.

NOTES

1 The Oxford cosmologist E.A. Milne postulated, as a 'cosmological principle', that the large-scale universe (about which hardly anything was then known) was uniform enough to make the simple theoretical models applicable. Bondi and Gold went further, basing their steady-state theory on a so-called 'perfect cosmological principle' – that the universe was the same *at all times* as well as at all places. We quite quickly learned that the latter was false (see Chapters 2 and 3). In contrast, Milne's 'cosmological principle' has been vindicated more accurately than he would have dared to hope, for the part of the universe within our present horizon, Mach's principle, described in Chapter 12, has been a more fruitful concept than other 'principles', though perhaps only because it has been interpreted in so many different ways.

2 Before Smolin's idea (first proposed in an article in the journal *Classical and Quantum Gravity*, vol. 9, p173, (1992)) can be developed further, two issues need clearer formulation. First, does the selection process favour universes that generate black holes at the *maximum rate* (and maximum efficiency), or is the *total number produced by a universe over its lifetime* more relevant? The latter criterion would depend, more than anything else, on how big and long-lived a universe was. The second point concerns the anthropic constraint. Suppose it turned out that black holes were most readily produced in a universe where complex life could never evolve. For instance, stars might more readily form black holes if there were no nuclear energy sources, and no stable elements other than hydrogen. But there would then be no chemistry, and perhaps no complexity. Should that be so, one might reject Smolin's ideas entirely. However one could then test a modified prediction: that our universe produces more black holes than any other *in which conditions are equally propitious for complex evolution*.

3 This quotation is from p359 of Heinz Pagels' book *Perfect Symmetry* (Michael Joseph, 1985).

4 This resembles the line of argument that could have been used against Boltzmann's hypothesis that our universe was a fluctuation away from equilibrium (see Chapter 13) – our existence requires an improbable fluctuation, but the actual universe is vastly larger and constitutes an immensely more improbable fluctuation than our existence demands.

FURTHER READING

J. Audouze and G. Israel (editors), *Cambridge Atlas of Astronomy* (Cambridge University Press)
J. D. Barrow, *Theories of Everything* (Oxford University Press)
J. D. Barrow and F. Tipler, *The Anthropic Cosmological Principle* (Oxford University Press)
P. Davis, *About Time* (Viking)
A. Dressler, *Voyage to the Great Attractor* (Doubleday)
F. J. Dyson, *Infinite in All Directions* (Penguin)
G. F. R. Ellis, *Before the Beginning: Cosmology Explained* (Boyars/ Bowerdean)
M. Gell-Mann, *The Quark and the Jaguar* (W. H. Freeman)
E. R. Harrison, *Masks of the Universe* (Macmillan)
S. W. Hawking, *A Brief History of Time* (Bantam)
S. W. Hawking and R. Penrose, *The Nature of Space and Time* (Princeton University Press)
F. Hoyle, *Home is Where the Wind Blows: Chapters from a Cosmologist's Life* (Mill Valley: University Science Books)
M. Kaku, *Hyperspace* (Oxford University Press)

L. Lederman and D. Schramm, *From Quarks to the Cosmos* (W. H. Freeman)

A. Lightman and R. Brewer, *Origins: the Lives and Worlds of Modern Cosmologists* (Harvard University Press)

A. Linde, *Particle Physics and Inflationary Cosmology* (Harwood)

M. S. Longair, *Our Evolving Universe* (Cambridge University Press)

S. P. Maran (editor), *Astronomy and Astrophysics Encyclopaedia* (Van Nostrand)

J. North, *History of Astronomy and Cosmology* (Fontana)

I. D. Novikov, *Black Holes and the Universe* (Cambridge University Press)

A. Pais, *Inward Bound: of Matter and Forces in the Physical World* (Oxford University Press)

A. Pais, *Subtle is the Lord: the Science and Life of Albert Einstein* (Oxford University Press)

P. J. E. Peebles, *Principles of Physical Cosmology* (Princeton University Press)

R. Penrose, *The Emperor's New Mind* (Oxford University Press)

M. Rees, *Perspectives in Astrophysical Cosmology* (Cambridge University Press)

M. Rowan-Robinson, *Ripples in the Cosmos* (W. H. Freeman)

C. Sagan, *Cosmos* (MacDonald)

F. Shu, *The Physical Universe* (Mill Valley: University Science Books)

J. Silk, *A Short History of the Universe* (W. H. Freeman)

K. S. Thorne, *Black Holes and Time Warps: Einstein's Outrageous Legacy* (Norton)

K. C. Wali, *Chandra: a Biography of S. Chandrasekhar* (Chicago University Press)

S. Weinberg, *Dreams of a Final Theory* (Pantheon)

S. Weinberg, *The First Three Minutes* (Basic Books)

J. A. Wheeler, *A Journey into Gravity and Spacetime* (W. H. Freeman)

F. Wilczek and B. Devine, *Longing for the Harmonies* (Norton)

INDEX